CATBox

Winfried Hochstättler · Alexander Schliep

CATBox

An Interactive Course in Combinatorial
Optimization

 Springer

Winfried Hochstättler
Fakultät für Mathematik und Informatik
FernUniversität in Hagen
58084 Hagen
Germany
Winfried.Hochstaettler@FernUni-Hagen.de

Alexander Schliep
Department of Computer Science and
BioMaPS Institute for Quantitative Biology
Rutgers, The State University of New Jersey
110 Frelinghuysen Road
Piscataway, NJ 08854-8019
USA

Additional material to this book can be downloaded from http://extra.springer.com.

ISBN 978-3-540-14887-6   e-ISBN 978-3-642-03822-8
DOI 10.1007/978-3-642-03822-8
Springer Heidelberg Dordrecht London New York

Library of Congress Control Number: 2009939264

Mathematics Subject Classification (2000): 97-01, 05C85, 68W40, 90C35, 90C27, 97U50, 05C38, 05C70

Printed on acid-free paper

Springer is part of Springer Science+Business Media (www.springer.com)

Preface

As naturally as static pictures were used to communicate mathematics from the very beginning, the advent of algorithmic and computational mathematics—and the availability of graphical workstations—introduced us to dynamic displays of mathematics at work. These animations of algorithms looked in the case of graph algorithms quite similar to what you see in CATBox. In fact there has been a substantial literature on the topic and software systems implementing such animations is obtainable from various sources. Nevertheless, these systems have not found very widespread use, with few notable exceptions.

Indeed, this incarnation of CATBox, both concept and name are due to Achim Bachem, was motivated by similar experiences with two prior projects under the same name. The difficulty in using these systems in teaching was that the animations were decoupled from algorithmic code or simply displayed pseudo code, with a hidden algorithm implementation doing its magic behind the scenes. More importantly, the algorithm animation system was almost always a separate entity from the textbook and they were developed respectively written by different people.

This lead to the idea of constructing a system where the student can control the flow of the algorithm in a debugger like fashion and what she or he sees is the actual code running, developed hand-in-hand with a course book explaining the ideas and the mathematics behind the algorithms. Our concept of course influenced our choices in problems and algorithms introduced.

We cover classical polynomial-time methods on graphs and networks from combinatorial optimization, suitable for a course for (advanced) undergraduate or graduate students. In most of the methods considered, linear programming duality plays a key role, sometimes more, sometimes less explicitly. In our belief some of the algorithmic concepts, in particular minimum-cost flow and weighted matching, cannot be fully understood without knowledge of linear programming duality. We decided to already present the simplest example from that point of view and interpret Kruskal's greedy algorithm to compute a minimum spanning tree as a primal-dual algorithm. For that purpose we have to introduce the basic ideas of polyhedral combinatorics quite early. This might be tedious for an audience mostly interested in shortest paths and maximum flows.

Material from this book has been used in full semester graduate courses and for sections of graduate courses and seminars by colleagues and us. For use in an undergraduate course, we suggest to skip the linear programming related Chaps. 4, 7 and 9 possibly augmenting them with material from other sources.

The animation system Gato is open source, licensed freely under the GNU Lesser General Public License (LGPL). We will collect algorithms which did not make it into the book and community contributions under a compatible license, to support teaching a wider range of graph algorithms in the future.

Acknowledgments

Many people have contributed at various stages to the book and software and were essential for the success. In particular we would like to thank our mentors and collaborators from the Center for Parallel Computing (ZPR) at the University of Cologne, Achim Bachem, Sándor Fekete, Christoph Moll and Rainer Schrader for their motivation and many helpful discussions. We would also like to thank Martin Vingron at the Max Planck institute for Molecular Genetics for support. Many thanks to Günther Rote for helpful suggestions and finding bugs in the manuscript and algorithm implementations. Benjamin Georgi, Ivan G. Costa, Wasinee Rungsarityotin, Ruben Schilling, Stephan Dominique Andres read the manuscript and helped to iron out the rough spots and reduce the number of errors.

The software Gato benefitted from experiences with a prior implementation provided by Bernd Stevens and Stefan Kromberg. Ramazan Buzdemir, Achim Gädke and Janne Grunau contributed greatly to Gato; Heidrun Krimmel provided the screen design. Torsten Pattberg provided implementations of many algorithm animations and helped to increase coherence between text and implementations. Many, many thanks for their great work and the fun time working together.

Many beta testers have been using CATBox and helped to find bugs and inconsistencies. In particular we would like to thank Ulrich Blasum, Mona K. Gavin, Dagmar Groth, Günther Rote, Wolfgang Lindner, Andrés Becerra Sandoval, Werner Nemecek, Philippe Fortemps, Martin Gruber, and Gerd Walther.

Last but not least, many thanks to Martin Peters, our editor at Springer and very likely the most patient man on earth, for his consistent support.

Hagen, Germany   Winfried Hochstättler
Piscataway, NJ, USA   Alexander Schliep
August 2009

Contents

Chapter 1
Discrete Problems from Applications

This book is about algorithms which solve some classical problems from discrete optimization. Such discrete problems arise in many application areas—from communications to scheduling in factories, from biology to economics. The field has attracted considerable attention over the last five decades mainly for one reason, namely the availability of powerful computing machinery. Let us motivate our topic with an example.

Application 1 (minimal spanning tree—MST) A company installs a new computer network. Computers on everybody's desk have to be connected to the server, possibly via some other computer. Establishing a network link between two machines costs a specific amount of money depending on their distance. The company is interested in keeping the total cost of installing the network low.

Clearly, the number of distinct networks of this kind is finite. Be aware, though, that a discrete problem does not always have a finite number of possible solutions.

One reason that questions like this have not been considered an important problem until the midst of the twentieth century is that they are intractable by hand except for very small problem instances. If the company had only, say, 3 computers you can simply check for all the possible 64 networks whether they connect every computer to the server, compute their costs and select the cheapest network among them. Unfortunately, for ten computers the number of possible networks is already 36,028,797,018,963,968. For twenty, the number of possibilities has 64 digits! To generate and check all possible networks is neither efficient nor feasible and we will learn about a much easier, efficient and systematic way of finding such a network later.

We agree that our model for a computer network does not reflect today's hardware technology, but this is not an issue. When we apply mathematical methods to problems, we try to solve the *abstract essence* of a problem. Sometimes we do have to neglect particularities of a problem, at least at first, to arrive at a solution. The advantage of this approach is that distinct problems from applications may lead to the same abstract problem which we then already have solved once for all. Furthermore, the real problem may not be solvable in an exact sense but we possibly obtain an approximative answer by using the mathematical method.

W. Hochstättler, A. Schliep, *CATBox*, DOI 10.1007/978-3-642-03822-8_1,
© Springer-Verlag Berlin Heidelberg 2010

We shall attempt to solve the problems we encounter by developing algorithms. An algorithm is a strict, formal recipe which tells us how to compute a solution to a particular class of problems. We will give a slightly more formal definition in the next section.

Software Exercise 2 Before doing so, let us have a break and study the problem of finding such a network with our software. In Chap. 3 we will study this problem in detail and call it Minimal Spanning Tree (MST). In Appendix A you can find a detailed introduction to using Gato. Starting the program Gato you will find yourself with two windows one of which contains a menu bar. Select in the Menu `Window Layout` the menu item `One Graph Window`.

Now, select in the Menu `File` the menu item `Open Algorithm`. From the directory `03-MinimalSpanningTrees` choose `MSTInteractive.alg`. Now the following ten lines of *Python* code, see Appendix B for a brief introduction to Python, will appear in the *Algorithm Window* to the left of your screen.

ALGORITHM MSTInteractive

```
1  while E.IsNotEmpty():

       (u,v) = PickEdge()

       if not SameComponent(u,v):
6          Merge(u,v)
       else:
           AddEdgeToComponent((u,v))
           Circuit = FindCircuit(u,component[u])

11         if Circuit != None:
               e = MaximalWeightEdge(Circuit)
               DeleteEdge(e,Circuit,component[u])
```

Select a graph for the algorithm to compute an MST on. Select in the Menu `File` the menu item `Open Graph` to load the graph `MinimalSpanningTrees08.cat`. When you are done, your screen should look similar to Fig. 1.1.

The Graph Window depicts a graph consisting of line segments, which we call edges, and numbered points, which we call vertices. How does the code in the "Algorithm Window" operate on this? While there are still edges which are not explored, we have to pick an edge—the line of code is displayed in green to indicate that some interaction from you is required—and examine whether this line connects two previously unconnected parts of the network. If so, we add it to our set of required connections, otherwise we check whether we can decrease the cost of the current connections by using that edge to connect vertices.

Click on the *Start*-Button at the bottom of the Algorithm Window. First, all vertices in the Graph Window get different colors to indicate that they are not connected. The algorithm starts and waits at the green line, the one containing the command `PickEdge()`, for you to pick an edge. Click at some edge in the Graph Window. It will change its color to the color of one of its endpoints, so will the other endpoint of the edge. You may now continue adding edges to your set of connections. Pay attention to the line of text at the bottom of the graph window. Unless the mouse pointer is located above a vertex or an edge you can see the cost

of the largest connected component of the present solution and also the cost of a
minimal connection of all points.

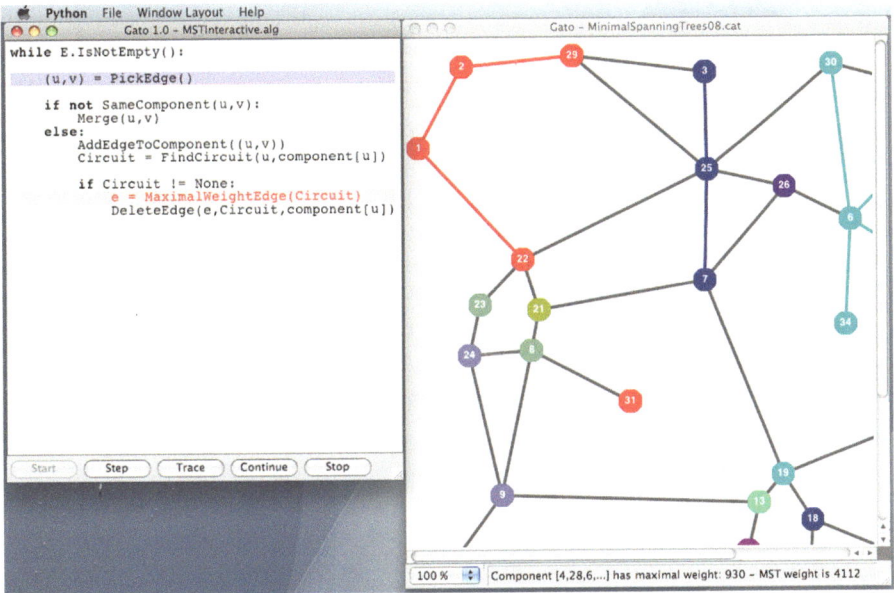

Fig. 1.1 We show Gato with the algorithm `MSTInteractive.alg` running on the graph
`MinimalSpanningTrees08.cat`

Continuing this way you might encounter a yellow closed curve—a *circuit*.
Mouse clicks in the graph window do not get you any further, because the red line
in the other window, we call it the *Debug Window*, indicates that a *break point* has
been set at the code line

```
e = MaximalWeightEdge(Circuit).
```

Click on the *Step*-Button. Now the algorithm moves to the next line

```
DeleteEdge(e,Circuit,component[u])
```

and one edge of the circuit in the Graph Window turns red. The red edge is the
longest edge in the circuit and its removal leaves its endpoints as well as the other
vertices in the circuit connected. Push the *Continue*-Button and continue picking
edges. If you close a circuit again you may repeat the above process.

To make life a bit more comfortable you may click on the red line which should
immediately turn grey, indicating that you have deleted the break point at this line.
The algorithm will subsequently run across this line without stopping. If you want
to add a breakpoint, just click on a grey line to turn it red.

The text line at the bottom of the Graph Window will tell when your tree is
spanning—i.e., whether it connects all the vertices—and how much it costs. By
working through the edges you cannot avoid to arrive at an optimal tree.

We present further sample problems from different areas that can be solved with the methods presented in this book.

Application 3 (shortest path) Given a network of roads with fixed, known lengths of road segments, determine a shortest path (a route on the road segments of the network) from a specified location to a destination.

Application 4 (shortest path) A DNA molecule is a long sequence of letters, a word, on the alphabet $\{A, C, G, T\}$. An important problem is to estimate the degree of relationship or the distance of two DNA-sequences, say TAGACAGA and CACGAGA. It is assumed that letters are inserted or deleted or mutate with a certain probability during evolution. Now, the similarity of two sequences under this model of evolution is the maximum of the product of the probabilities of a sequence of operations that transforms one sequence into the other.

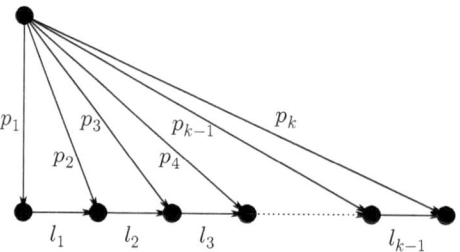

Fig. 1.2 The network for the lot sizing problem without intermediate products in Application 5. The lot sizes correspond to the flow through the edges with edge cost p_i; the demands d_i determine the required in-flow of the k vertices at the bottom

Application 5 (shortest path, mincost-flow) The demand of a product for k periods has been estimated to be $d_1, \ldots, d_k$. The production cost varies with the period; in period j a unit of the product is manufactured for a cost of p_j. The cost for stock-keeping of a unit of the product from period $j - 1$ to period j is l_{j-1}. What lot sizes shall be used to produce the good? Often there are additional capacity constraints on the lot sizes (Fig. 1.2).

If the good is processed in several steps with several intermediate products that can be held in stock as well we get a more complicated problem with the following data and variables:

- d_j, the demand in period j,
- p_{kj}, the cost of a unit in step k of production in period j, and
- l_{kj}, the cost of stock keeping of a unit of k-th intermediate product in period j,

as well as capacity constraints.

Application 6 (weighted bipartite matching) The objective in chromosome karyotyping is to map the 46 chromosomes to the 23 chromosome classes. Membership of chromosome i to class j is given with a probability p_{ij}. This data may be for example the output of some pattern recognition program. Find the matching of highest probability.

Application 7 (shortest path and weighted matching) A postal worker is supposed to deliver mail in a residential area in such a way that all streets are traversed and the total length of his delivery route is minimized.

Chapter 2
Basics, Notation and Data Structures

In the example that we used to introduce the software and in all our other examples the fundamental objects are graphs, that is a set of objects and links connecting or relating those objects. Sometimes the links bear additional data, like a length or a capacity and sometimes they are directed. Often, graphs are introduced as sets of vertices, the objects, and sets of edges, the links, where an edge is either an ordered or unordered pair of vertices. We prefer to proceed a bit more formally. The following definition allows to distinguish different links between a single pair of objects. What we will define here as a *graph* is sometimes also called a *multigraph*.

2.1 Definition of a Graph

Definition 1 (Digraph, Graph) Let V be a finite set (of *vertices*) and A a finite set (of *arcs*). We define the following two maps for A.

$$\text{tail} : A \rightarrow V \qquad\qquad \text{head} : A \rightarrow V$$
$$a \mapsto \text{tail}(a), \qquad\qquad a \mapsto \text{head}(a)$$

They map every arc to a unique beginning respectively end which we call *tail* and *head* respectively. If a vertex v is tail or head of an arc a, we say that v is *incident* to a. The quadruple $D = (V, A, \text{tail}, \text{head})$ is called a *directed graph* or shortly *digraph*.

If a direction of an arc does not matter we call the arc an *edge* and denote the set of edges by E. This way we get a triple $G = (V, E, \varphi)$, where φ is the *incidence function* mapping each edge to a set of one or two vertices (the end vertices of that edge). Such a triple is named a *graph*. Often φ is omitted and we write for short $G = (V, E)$.

If head and tail coincide, that is, an edge is incident to only one vertex, then we call it a *loop*. To simplify notation we frequently denote an edge e in a graph as a set of end vertices, $e = \{v_1, v_2\}$. Similarly, for a digraph $D = (V, A)$ we denote an arc $a \in A$ as the ordered pair (v_1, v_2) if, and only if, $\text{tail}(a) = v_1$ and $\text{head}(a) = v_2$.

W. Hochstättler, A. Schliep, *CATBox*, DOI 10.1007/978-3-642-03822-8_2,
© Springer-Verlag Berlin Heidelberg 2010

Slightly abusing notation, as it is common in the literature, we will frequently denote an edge $\{v_1, v_2\}$ in a graph by (v_1, v_2) although the graph is not oriented.

Example 1 Figure 2.1 on the left side depicts a digraph with vertex set $\{1, 2, 3\}$ and arcs $\{a, b, c, d\}$ and the following maps tail and head

arc	head	tail
a	1	2
b	1	3
c	2	3
d	3	3

and on the right side a graph with vertex set $\{A, B, C\}$ and edge set $\{w, x, y, z\}$ and incidence function

	x	y	z	w
$\varphi(\cdot)$	$\{A, B\}$	$\{A, C\}$	$\{B, C\}$	$\{C\}$

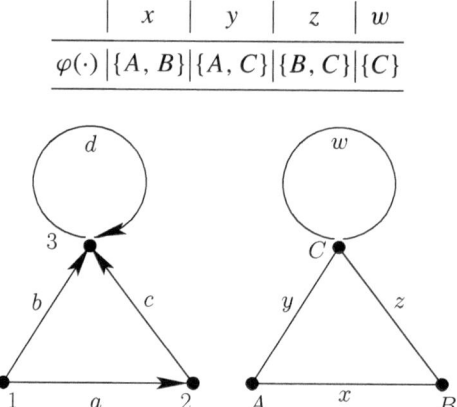

Fig. 2.1 A digraph (*left*) and a graph (*right*)

Definition 2 Let $G = (V, E)$ be a graph and $v \in V$. The number of edges incident with v,

$$\deg_G(v) := |\{e \in E \mid v \in \varphi(e)\}| + |\{e \in E \mid \varphi(e) = \{v\}\}|,$$

$\deg_G(v)$ is called the *degree of v in G*. We will frequently omit the subscript G if only one graph is involved. Note, that loops in an undirected graph are counted twice for the vertex degree. Similarly, we define outdegree $\deg_D^+$ and indegree $\deg D^-$ of a digraph D as the number of outgoing respectively incoming arcs of the vertices.

Consider the graphs in Fig. 2.1. Vertex C has a degree of $\deg(C) = 4$, while $\deg(A) = \deg(B) = 2$. Vertex 3 has outdegree $\deg^+(3) = 1$ and indegree $\deg^-(3) = 3$.

Exercise 8 There is a graph editor called Gred included with CATBox. How to use it is explained in Appendix A.4. When you start Gred, clicking on the empty canvas will add a vertex at the position where you clicked. If you select the *Add edge tool* by clicking on the icon depicting an edge, you can add edges between vertices by clicking and dragging the mouse. Try to create a triangle and a directed triangle, i.e. the graph and the digraph in Fig. 2.1 without the loops with Gred and save it as `gred.cat`.

2.2 How to Represent a Graph on the Computer?

A classical simple method to represent a digraph is to store its *vertex edge incidence matrix B*. Here B is a $(|V| \times |A|)$-matrix, where each arc $a = (h, t)$ is represented by a column. The entries of this column are zero except for a 1 in the h-th row and a -1 in the t-th row; note, that this precludes existence of loops. With graphs we proceed in an analogous fashion. The matrices for the graph and the digraph from our example shown in Fig. 2.1 are

$$
\begin{array}{c}
\quad a \quad b \quad c \quad d \\
\begin{array}{c}1\\2\\3\end{array}
\left(\begin{array}{cccc}
1 & 1 & 0 & 0 \\
-1 & 0 & 1 & 0 \\
0 & -1 & -1 & ?
\end{array}\right)
\end{array}
, \text{ and }
\begin{array}{c}
\quad x \quad y \quad z \quad w \\
\begin{array}{c}A\\B\\C\end{array}
\left(\begin{array}{cccc}
1 & 1 & 0 & 0 \\
1 & 0 & 1 & 0 \\
0 & 1 & 1 & ?
\end{array}\right).
\end{array}
$$

Another possibility to store a (di)graph as a matrix is the *adjacency matrix*. We say that two vertices are *adjacent*, if they are joined by an arc (an edge). The adjacency matrix C of a (di)graph is the $(|V| \times |V|)$-matrix, where c_{uv} is the number of arcs from u to v, respectively, of edges between u and v:

$$
\begin{array}{c}
\quad 1 \quad 2 \quad 3 \\
\begin{array}{c}1\\2\\3\end{array}
\left(\begin{array}{ccc}
0 & 1 & 1 \\
0 & 0 & 1 \\
0 & 0 & 1
\end{array}\right)
\end{array}
\qquad
\begin{array}{c}
\quad A \quad B \quad C \\
\begin{array}{c}A\\B\\C\end{array}
\left(\begin{array}{ccc}
0 & 1 & 1 \\
1 & 0 & 1 \\
1 & 1 & 1
\end{array}\right).
\end{array}
$$

In applications, graphs have typically many fewer edges than maximally possible; in that case, they are called *sparse*. For sparse graphs the incidence matrix respectively adjacency matrix will mostly contain zeroes. Nevertheless, the incidence matrix requires $|V| \cdot |E|$ and the adjacency matrix $|V|^2$ entries. To estimate the memory necessary to store that data we assume that a vertex or an edge requires a constant amount of the memory. Then an incidence matrix requires $O(|V| \cdot |E|)$ and an adjacency matrix $O(|V|^2)$ space. Here we used a Landau-symbol or the so called "big-Oh"-notation which is defined as follows:

If $f, g: \mathbb{N} \to \mathbb{N}$ are functions then $g \in O(f)$ if $\exists n_0 \in \mathbb{N}, \exists c \in \mathbb{N}, \ \forall n \geq n_0: g(n) \leq cf(n)$. That is, there is a constant c and an integer n_0 such that for all large enough integers $g(n) \leq cf(n)$.

There is a more efficient way to encode a graph by storing a list of arcs instead. It is frequently necessary in the algorithms to access all arcs incident with a vertex and the vertices incident with an arc. This can be done with a data structure based on (doubly) linked lists. In Fig. 2.2 we sketch the relevant information we need to store. For fast access we keep pointers to the vertices in an array. Each vertex v is equipped with a linked list of pointers to the arcs emanating from it and in case of a digraph a second pointer to a list of those ending in v. Vice versa, each arc has pointers to its endpoints.

This has the advantage of fast access, a memory requirement of $O(|V|+|E|)$ and high flexibility. It is, for example, easy to store additional information about vertices (say a demand) or arcs (a length or capacity).

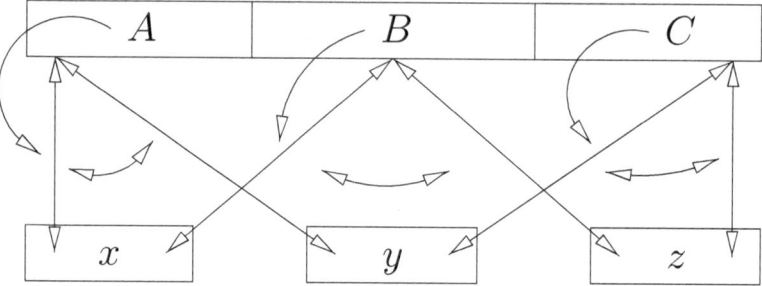

Fig. 2.2 A data structure based on linked lists of vertices and incident edges which allows efficient traversals. For simplicity pointers to edges and to vertices are displayed as one line with double *arrows*. The (doubly) linked list for transversal of incident edges are shown in a circular shape

Remark 1 The presented data structure is chosen to give best results for a theoretical analysis. The order of the running time does not change with an additional pointer for each vertex or edge. For practically efficient implementations of graph algorithms for a specific purpose you should consult the latest literature. The methods and data structures necessary for efficient computations after all depend on the current state of the art of hard- and software. A further aspect which should influence your decisions is whether the graphs are static or dynamic, i.e. change at runtime.

2.3 Basic Terminology from Graph Theory

Definition 3 Let $D = (V, A)$ be a digraph. We call an alternating sequence $v_1 a_1 v_2 a_2 v_3 \ldots a_n v_{n+1}$ of vertices and arcs a *chain*, if for all $i = 1, \ldots, n$

$$\text{tail}(a_i) = v_i \quad \text{and} \quad \text{head}(a_i) = v_{i+1}. \tag{2.1}$$

A chain is a *cycle* if $v_1 = v_{n+1}$. A chain is called a *path* if no vertex is visited twice, i.e. $1 \le i < j \le n + 1 \Rightarrow v_i \ne v_j$. A cycle $v_1 a_1 v_2 a_2 v_3 \ldots a_n v_{n+1}$ is a *circuit*, if $v_1 a_1 v_2 a_2 v_3 \ldots a_{n-1} v_n$ is a path.

We shall frequently denote chains, cycles, paths and circuits just by one of their sequences of vertices respectively edges; e.g. we have a chain $v_1, v_2, \ldots, v_{n+1}$ or $a_1, \ldots, a_n$.

The same notions are defined for graphs as well by replacing arcs by edges and (2.1) with $\varphi(e_i) = \{v_i, v_{i+1}\}$. In the following we will no longer mention explicitly how to transfer a definition from a digraph to a graph whenever it is straight-forward.

Definition 4 Let $D = (V, A)$ be a digraph, $W \subseteq V$ a subset of its vertex set and $B \subseteq A$ a subset of its arc set, such that the restriction $tail_{|B}$ of $tail$ and $head_{|B}$ of $head$ to B map to W. Then the digraph $H = (W, B, head_{|B}, tail_{|B})$ is called a *sub-digraph* of D.

Definition 5 Let $G = (V, E)$ be a graph. If for each pair of vertices v, w there is a path starting in v and ending in w, we call G *connected*. Let $D = (V, A)$ be a digraph. D is called *strongly connected*, if there is a directed path starting in v and ending in w for any ordered pair of vertices v, w. That is, both directed paths from v to w and from w to v exist. The maximal (strongly) connected subgraphs of a graph G (respectively a digraph D), are called *(strong) components*, here maximal is understood with respect to edge set inclusion.

2.4 Algorithms and Complexity

We will not give a mathematically rigorous definition of an algorithm and refer to books on computer science or computational complexity theory, for example [21], for that purpose. We will give an informal notion of an algorithm instead and introduce the *worst case complexity* of an algorithm by an example. An algorithm is an exact sequence of instructions for computations or for data processing. By exact we mean that an algorithm

(i)   is described in a *finite text*,
(ii) the *sequence* of the steps is *unique*[1] in any computation,
(iii) each *step* of the procedure can be executed *mechanically and efficiently*, and
(iv) that the algorithm *terminates* after a finite number of steps.

The language used in CATBox is *Python*. It enables us to denote the algorithms in an abstract way comparable to *pseudo-code* but with the added benefit that the algorithm can actually be executed. Let us demonstrate what we mean by an algorithm using the computation of the components of a graph as an example. We assume that the graph is given as a list of `Vertices` where each vertex v **in** `Vertices` is equipped with an adjacency list `Neighborhood(v)`.

---

[1] This condition can be relaxed, allowing for randomized sequences in an algorithm. The probabilistic techniques for analysis of such *randomized algorithms* and an introduction can, for example, be found in [32].

One possible algorithmic approach would be to use the definition and check whether its conditions are satisfied. For that purpose we had to examine all sub-graphs for being connected. For the latter we might search for a path for each pair of vertices. Finally, from the connected subgraphs we had to find those which are maximal with respect to inclusion. This algorithm which naively implements the definition has a major drawback. While the algorithm will terminate after finitely many steps, we would have to wait forever, at least longer than our life span, for the algorithm to stop already for graphs of modest size. For a graph with vertex set V there are $2^{|V|}$ subsets of vertices and for each of them we will have at least one subgraph. For a graph with 1,000 vertices this would amount to more than 10^{30} vertex sets.

The following presents a much more efficient alternative:

ALGORITHM BFS-components

```
step = 1

for root in Vertices:
    if not label[root]:
        component[root] = NewComponent()
        component[root].AddVertex(root)
        Q.Append(root)
        pred[root] = root

        while Q.IsNotEmpty():
            v = Q.Top()
            label[v] = step
            step = step + 1
            for w in Neighborhood(v):
                if not pred[w]:
                    component[v].AddVertex(w)
                    component[w] = component[v]
                    Q.Append(w)
                    pred[w] = v
```

Every vertex w has a label `component[w]`, indicating the component that w belongs to and a pointer to its *predecessor* `pred[w]`. In the beginning no vertex has a predecessor, that is `pred[w]` = None for all w. We start with the first unprocessed vertex v in the list and compute the unique component that this vertex belongs to, for example by computing a BFS-tree (breadth first search). Such graph traversal algorithms are discussed below.

Thus, we keep a queue Q, i.e. a first-in-first-out data structure, that is initialized with v. We proceed by removing the next vertex `Q.top()` from the queue, mark all its unmarked neighbors w in `Neighborhood(v)` as belonging to the component containing v, store the edge $e = (pred[w], w)$ as a proof of connectedness and add w to the queue.

Once the queue Q is empty we search for the next unprocessed vertex and continue the search in the, thus, newly discovered component.

Software Exercise 9 Start Gato and load the algorithm `BFS-components.alg` from the directory `02-Graphs and Networks`. Do not forget to load a graph as well, we suggest to explore `3Components.cat`. You will find it and other examples in the same directory. Now you can follow the algorithm step by step or explore its dynamics.

After having defined the algorithm above we should convince ourselves of its *correctness*. Therefore, we need to prove that it computes the connected components. Furthermore we should analyze its *running time behavior*.

2.4.1 Correctness

Although the algorithm is intuitively clear, our exact proof of its correctness is tedious. Nevertheless, since intuition sometimes fails, a proof should be mandatory for an algorithm.

Proposition 1 *(i)* `BFS-components.alg` *does always terminate.*
(ii) *When the algorithm terminates every vertex has received a label* `comp-onent[v]` *and a predecessor* `pred[v]` *different from* `None`.
(iii) *If* $w \in$ `Vertices` *is a vertex and* `component[w]` $=$ `component[v]`, *then there exists a chain from* w *to* v *in* G.
(iv) *If there exists a chain from* w *to* v *in* G *then* `component[w]` $=$ `comp-onent[v]`.

Proof   The only critical statement for finiteness is the **while**-loop. Hence, it suffices to verify that each vertex is appended to Q exactly once to prove the first claim. Since a vertex is added to the queue only if its predecessor is changed from `None` to a well defined value and this value is never changed in the following, finiteness follows. We check in line (15) for each vertex v, whether its predecessor `pred` is still `None` and, if so, change it and give it a label. Thus the second claim follows.

Now let `component[w]` $= v_0$. We prove the third statement by induction over the steps of the algorithm. If $w = v_0$, then the path v_0 is a $v_0 - w$-path, otherwise w has received its label in line (17) of the algorithm from some neighbor v. Vertex v thus must have been labeled at an earlier time. By the inductive assumption there is a v_0v-chain, which can be extended to a v_0w-chain by the edge (vw).

For the last claim we prove by induction over the algorithm that the vertices which are processed in each execution of the inner **while**-loop all receive the same label `component[w]` $=$ `component[root]`. When the **while**-loop is entered the queue consists solely of `root`. Thus, all of `roots` neighbors are added to the `component` of `root`. Clearly, applying inductive assumption in later iterations immediately yields the assertion.

It is left to verify, that if there exists a chain from v to w then the vertices are processed in the same execution of the inner **while**-loop. This is done by induction on the length k of the chain. First, we consider the case $k = 1$. We may assume that v has entered the queue before w did. All vertices that enter the queue until v is processed in line (11) receive the same label as v. Thus, when the neighborhood of v is explored in lines (14) $-$ (19) w is assigned the label `component[v]`, unless it already bears a label, necessarily `component[v]`. Now if $k > 1$ let v_1 denote the first vertex on the vw-chain. By induction we have `component[v]` $=$ `component[$v_1$]` $=$ `component[w]`.   □

2.4.2 Running Time

Let us analyze the running time of the method. For that we will count the number of *elementary operations* that are executed by the procedure until termination. We will use a generous definition of elementary operation, as we are not interested in quantifying the running time in minutes and seconds on a particular computer. Rather, we would like a qualitative statement concerning the *order* of the running time in relation to the *size* of the input. Instead of absolute—or wall-clock times— we are content with knowing whether an algorithm will run twice, four or eight times as long when we double the size of the input. The respective complexities are called *linear*, denoted by $O(n)$, *quadratic* $O(n^2)$, or *cubic* $O(n^3)$. Therefore, we will consider any computation that can be performed in constant time as an elementary operation. In particular, we will ignore fixed, scalar multiples of the number of elementary operations, which allows us to simply count each line of the algorithm as performing an elementary operation.

The first loop is executed $|V|$ times, so is the second. In each loop a constant number of operations is performed. As seen in the proof of Prop. 1 every vertex v is added to Q at most once. Also, we process each vertex in the **while**-loop at most once. As a consequence, the second **for**-loop is executed at most twice for each edge. Summarizing we obtain:

Proposition 2 *The running time of BFS is $O(|V| + |E|)$.*

We say that the algorithm runs in *linear time*. We also usually set $m = |E|$ and $n = |V|$ and thus can write the running time as $O(n + m)$. Note, in our analysis we considered only upper bounds for the running time. This is common and is called *worst case* analysis. It gives a guarantee for the maximal number of operations needed for *any* input to compute a solution. Other counting methods like average running time play a minor role in theory, although there are applications where you would prefer algorithms which are faster for an average problem. In our case the running time must at least be $O(n + m)$, since every vertex and edge needs to be processed at least once.

Remark 2 The set of edges $\{(w, pred(w)) \mid w \in \text{Vertices}\}$ forms a *spanning forest*, see Definition 6 and Proposition 3 (on p. 20).

2.5 Two Graph Traversals

Here we will discuss the two most important methods to explore or traverse a connected graph, breadth-first-search (BFS) and depth-first-search (DFS). Given a connected graph $G = (V, E)$ the task is to visit each vertex exactly once in a most efficient way, starting from a vertex r which will be called the *root*. The algorithms can be used to order vertices according to the order visited or they are used as basic building blocks for more complicated algorithms, where further algorithmic steps are performed at different points in BFS or DFS.

In a *depth-first-search (DFS)* we always proceed to the first neighbor of a vertex that we encounter until we find that this already has been visited. If that happens, we proceed with the second neighbor and so on, until some vertex v has no unvisited neighbors. Then we step back to the vertex preceding v and continue. A DFS can easily be implemented recursively.

ALGORITHM DFS-Recursive

```
step = 1

def DFS(v):
    global step
5   label[v] = step
    step = step + 1
    for w in Neighborhood(v):
        if not label[w]:
            pred[w] = v
10          DFS(w)

root = PickVertex()
DFS(root)
```

You may follow the algorithm by applying it to the graph BFS.cat. We define a function DFS for a vertex v. We number the vertex with the next—globally known—free label step and start to explore its neighbors. When we find a neighbor w which has not received a number yet, we link it to v and call the function DFS for w. Once the algorithm is completed, we have labeled all elements reachable from the root.

When we analyze the complexity of this method it helps to realize that any edge is touched exactly twice in line 7 since, anytime we touch an edge either both end vertices are already numbered or we add a number to the second one while the first already bears such a label. Thus, we have a complexity of $O(|E| + |V|)$ as there may be less edges than vertices.

In a *breadth-first-search (BFS)* we first explore all neighbors of r and, in a next step, choose one of the newly visited vertices and explore their neighbors.

To implement BFS for a connected graph we have to keep an appropriate data structure. We use a *queue* object (see Appendix B), that is a data structure where we may *append* elements at its end and read and remove elements from its beginning. The latter is done by the method Top(). Now the remainder is easy, basically BFS-components.alg without the outer **for**-loop.

ALGORITHM BFS

```
root = PickVertex()
Q.Append(root)
pred[root] = root
step = 1
5
while Q.IsNotEmpty():
    v = Q.Top()
    label[v] = step
    step = step + 1
10  for w in Neighborhood(v):
        if not pred[w]:
            Q.Append(w)
            pred[w] = v
```

At first we put the specified vertex `root` into the queue. Then, as long as the queue is non-empty, we remove the first vertex v from the queue, link all its unnumbered neighbors to v and append them to the queue. A similar argument to Proposition 1 shows that again the running time is $O(|E| + |V|)$. Note, that we have an iterative, or non-recursive, implementation of BFS due to the queue.

A similar data structure is a *stack*. The only difference is that the data is added and removed at and from the top. It is a first-in-last-out data structure. The corresponding operations are `Push()` that puts an element on top of the stack and `Pop()` that removes the first element.

Exits

One of the first historical, or at least mythological, applications of the DFS was the exploration of the maze which was the habitat of the Minotaur to which King Minos of the island of Crete intended to sacrifice Theseus. Using advanced algorithmic skills and executing the algorithm himself, he used a thread bestowed upon him by the king's daughter, Ariadne, to backtrack and chalk marks to label areas visited. Besides maze exploration, motivated by the desire to find a way out, there are other simple graph traversals with important applications. The arcs of a digraph for example can be interpreted as implying an order relation between incident vertices, that is if $(i, j) \in A$ then $i < j$. Topological sorting [11], an extension of the DFS algorithm (cf. Exercise 14), can be used to return vertices in order or indicate that no such ordering exists. The latter is the case if the graph contains directed circuits (note that in algorithmic graph theory circuits are often referred to as cycles); that is a topological order only exists for directed acyclic graphs (DAG). Another application of the DFS is the computation of strongly connected components (SCC) [11], maximal subsets of vertices in a digraph such that every pair of vertices lies on a directed circuit.

Exercises

Exercise 10 Replace the BFS by a DFS (depth-first-search) by interchanging the queue with a *stack* S and the operations top() and append() with `S.push()` and `S.pop()` and as few other changes as possible. Operation push puts an element on top of the stack and pop removes the element on top. Should an element that has not yet been popped but resides somewhere in the stack ever be put on top of the stack?

Exercise 11 How can you modify the BFS-algorithm, such that it does not build a complete BFS-tree from some root node, but rather a tree of depth at most k containing only vertices of distance at most k from the root?

Exercise 12 We define the *girth* of a graph as the length of its shortest circuit. Develop an algorithm, which:

(i) Finds the shortest cycle through a given vertex.
(ii) Finds the girth of a graph.
(iii) Finds the girth of a bipartite graph. A graph G is called bipartite if the set of vertices V can be partitioned into V_1 and V_2, such that there are no edges e with both endpoints in the same partition.

Hint: Can you detect circuits with the BFS or DFS algorithm?

Exercise 13 What do BFS and DFS compute when the input is a directed graph? How can you detect directed circuits using one of these algorithms?

Exercise 14 Write an algorithm that, given an directed acyclic graph $D = (V, A)$, a DAG, computes an ordering $\sigma : V \to \{0, \ldots, |V| - 1\}$ such that for all $a = (u, v) \in A : \sigma(u) < \sigma(v)$.

Chapter 3
Minimum Spanning Trees

3.1 Minimum Connected Subgraphs

Recall the problem of installing a new computer network we introduced in Chap. 1. Using the vocabulary we have learned we can restate it more formally and more abstractly as the problem to find a *Minimum Connected Subgraph*:

Problem 1 (MCS) Let $G = (V, E)$ be a connected graph and $w : E \to \mathbb{Z}$ a *weight* or *cost function* on the edges. Find a connected subgraph $H = (V, F)$, such that

$$\sum_{e \in F} w(e)$$

is minimal.

As E has $2^{|E|}$ subsets finding a solution by enumeration is infeasible in practice, for all but the smallest graphs. We will develop a more efficient way of approaching this problem which can also handle large instances. To get an idea we first experiment with the CATBox software.

Software Exercise 15 Consider the example in Fig. 3.1, start the Software and load the algorithm `MSTInteractive.alg` and the graph `MinimalSpanning Trees09.cat` from the directory `03-MinimumSpanningTrees`.

What could be a reasonable approach? Try to start with an empty graph and successively add the cheapest edges as long as necessary. That is, as long as the graph is not connected.

Clearly, this approach will encourage you to add some unnecessary edges that are rejected by our interactive algorithm since they close a circuit with edges already included in our subgraph.

We can show that a solution does not contain a circuit as long as the edge weights are nonnegative.

Lemma 1 *Let $G = (V, E)$ be a connected graph and $w : E \to \mathbb{Z}_+$ a nonnegative edge weight function. Then there exists a solution F of Problem 1, that does not contain a circuit.*

W. Hochstättler, A. Schliep, *CATBox*, DOI 10.1007/978-3-642-03822-8_3,
© Springer-Verlag Berlin Heidelberg 2010

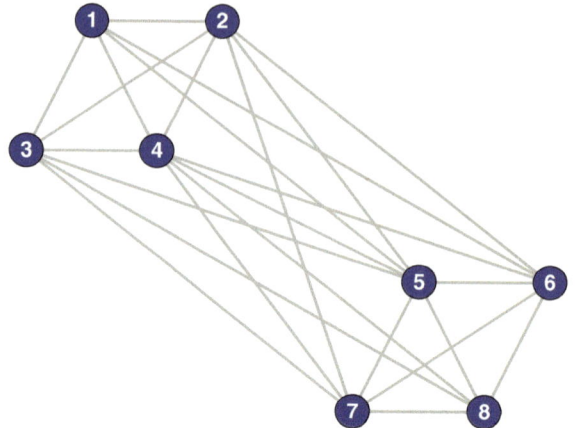

Fig. 3.1 The connected network from the file `MinimalSpanningTrees09.cat`

Proof We will prove this by contradiction. Let F be a solution that uses the least possible number of edges, that is for which $|F|$ is minimum. Assume F contains a circuit $e_1, \ldots, e_k$. Then it is possible to replace e_1 depending on the direction we traverse e_1 by $e_2, \ldots, e_k$ or $e_k, e_{k-1}, \ldots, e_2$ in any chain in F that uses e_1, thus obtaining a chain with the same end points. Hence $\widetilde{H} = (V, F \setminus e_1)$ is connected. Since F is a solution for Problem 1 we must have

$$\sum_{e \in F} w(e) \leq \sum_{e \in F \setminus \{e_1\}} w(e).$$

Thus, $w(e_1) \leq 0$ and since w is nonnegative, necessarily $w(e) = 0$ must hold. However, then $F \setminus \{e_1\}$ is a solution as well, contradicting minimality of $|F|$. $\qquad\square$

3.2 Trees

Connected graphs without a circuit are called trees and several trees make up a forest.

Definition 6 Let $G = (V, T)$ be a graph. Then G is called a *forest*, if T does not contain a circuit. A connected forest is called a *tree*.

Let $G = (V, E)$ be a graph. A forest $G_F = (V, F)$, where $F \subseteq E$, *spans* G or *is spanning for* G, if any two vertices u, v that are connected via a connecting path in G are connected in G_F as well.

Let $G = (V, T)$ be a tree and $v \in V$. Then v is a leaf of G, if $\deg(v) = 1$.

Lemma 2 *Every tree T with at least two vertices has at least two leaves.*

Proof Let u and v be the end points of a longest path in T. Clearly, u and v must be leaves; otherwise we could extend the path, since any edge to a vertex on the path other than its neighbor would close a circuit. $\qquad\square$

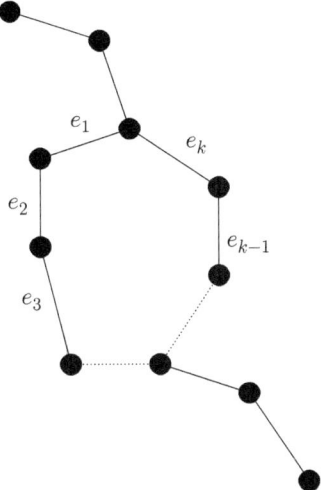

Fig. 3.2 Sketch for the proof of Lemma 1. Any chain using e_1 can be modified to use the edges $e_2, \ldots, e_k$ instead

Proposition 3 *Let $G = (V, T)$ be a graph. Then the following statements are pairwise equivalent:*

  (i) G is a tree.
 (ii) G is connected and $|T| = |V| - 1$.
(iii) G does not have a circuit and $|T| = |V| - 1$.
(iv) G has no loops and any two vertices are connected by a unique path.

Proof We proceed by induction on $|V|$, the assertion is trivial if $|V| + |T| = 1$, thus let $|V| + |T| \geq 2$.

$(i) \Rightarrow (ii)$ Let G be a tree. As G has no loops, we have $|V| \geq 2$, let v be a leaf of G. By definition G has to be connected. Let $G \setminus v$ denote the graph that arises from G by removing vertex v and the unique edge e incident to it. The resulting graph, clearly is still a tree and has one edge and one vertex less than G. Thus, the claim follows by induction.

$(ii) \Rightarrow (iii)$ We have to show that G has no circuit. Assume for a contradiction that G is connected, $|T| = |V| - 1$ and G has a circuit. Then, we iteratively remove an edge from each remaining circuit, always keeping the graph connected, until no more circuits exist. This way we find a connected graph $\widetilde{G} = (V, \widetilde{E})$ without circuit, hence a tree with $|\widetilde{E}| < |V| - 1$ contradicting what we have proven in $(i) \Rightarrow (ii)$.

$(iii) \Rightarrow (iv)$ Let $v, w \in V$. If v, w are connected by two different paths, then the symmetric difference of their edge sets contains a circuit, thus, by assumption we have at most one path. Let $G_1, \ldots, G_k$ denote the components of G.

Clearly, each component has no circuit and, thus, is a tree. By induction we calculate the number of edges in T to be $|V| - k$. Hence, we must have $k = 1$ and at least one path between any two vertices.

$(iv) \Rightarrow (i)$ By assumption G is connected. If G had a circuit we would get two vertices that are connected by different paths. Thus G must be a tree. □

3.3 Minimum Spanning Trees

We have seen that for non-negative weight functions there is always an optimal solution to Problem 1 that does not contain a circuit. Hence, we can formulate the following Problem 2. We shall see later that the more general case of Problem 1 easily reduces to it.

Problem 2 (MST) Let $G = (V, E)$ be a connected graph and $w : E \rightarrow \mathbb{Z}$ a weight function on the edges. Find a spanning tree T of G, such that

$$\sum_{e \in T} w(e)$$

is minimal.

Let us have a closer look at optimal solutions of this problem.

Lemma 3 *Let $G = (V, E, w)$ be an instance of Problem 2 and T an optimal solution. Then the following two criteria hold.*

Circuit criterion: *If $f = (s, t) \in E \setminus T$ and $e_1, \ldots, e_k$ is the unique path from s to t in T, then*

$$\forall i = 1, \ldots, k : w(e_i) \leq w(f).$$

Cut criterion: *For all $f \in T$ and all $g \in E \setminus T$ such that $(T \setminus \{f\}) \cup \{g\}$ is a tree, we have $w(f) \leq w(g)$.*

The first condition says that, in an optimal tree T, any edge f that does not belong to T is an edge of largest weight in the unique circuit in $T \cup \{f\}$. The cut criterion says that if $f \in T$ is an edge in an optimal solution, then it is an edge of minimum weight connecting the two components of $T \setminus \{f\}$.

Proof By Proposition 3 there is a unique path $e_1, \ldots, e_k$ from s to t in any tree. Assume we had some i such that $w(f) < w(e_i)$. Then $\widetilde{T} := (T \cup \{f\}) \setminus \{e_i\}$ is a tree (by Proposition 3) and $w(\widetilde{T}) < w(T)$ contradicting optimality of T. The second assertion is proved in a similar fashion. □

Actually, the above conditions are not only *necessary* but also *sufficient*, which will be shown in the following lemma.

Lemma 4 *Let* $G = (V, E, w)$ *be an instance of Problem 2. Then a tree* $T \subseteq E$ *is an optimal solution if and only if either one of the two equivalent conditions of Lemma 3 is satisfied.*

Proof First we show sufficiency of the circuit criterion by contradiction. Let T be a tree satisfying the circuit criterion and assume $\widetilde{T}$ were an optimal solution of Problem 2 and $w(\widetilde{T}) := \sum_{e \in \widetilde{T}} w(e) < w(T)$. Assume that $\widetilde{T}$ furthermore is chosen to be most similar to T, that is $|T \cap \widetilde{T}|$ is as large as possible. Let

$$f = (s, t) \in \widetilde{T} \setminus T \text{ denote an edge of smallest possible weight.} \qquad (3.1)$$

By assumption

$$w(e_i) \leq w(f) \qquad (3.2)$$

holds for all e_i on the s-t-path $e_1, \ldots, e_k$ in T. As $\widetilde{T}$ does not have a circuit, there is some i_0 such that $e_{i_0} = (u, v) \notin \widetilde{T}$ (see Fig. 3.3). Since $\widetilde{T}$ is optimal, the u-v-path $g_1, \ldots, g_l$ in $\widetilde{T}$ satisfies

$$w(g_j) \leq w(e_{i_0}) \qquad (3.3)$$

for all j. Clearly, there must be some $g_{j_0} \in \widetilde{T} \setminus T$, possibly $g_{j_0} = f$. As f was chosen with minimum weight putting all the inequalities together we have

$$w(f) \overset{(3.2)}{\geq} w(e_{i_0}) \overset{(3.3)}{\geq} w(g_{j_0}) \overset{(3.1)}{\geq} w(f),$$

thus they are all of the same weight. On the other hand $\widehat{T} := (\widetilde{T} \setminus \{g_{j_0}\}) \cup \{e_{i_0}\}$ is a spanning tree and $w(\widehat{T}) = w(\widetilde{T}) + w(e_{i_0}) - w(g_{j_0}) = w(\widetilde{T})$ and $|\widehat{T} \cap T| > |\widetilde{T} \cap T|$ contradicting the choice of $\widetilde{T}$.

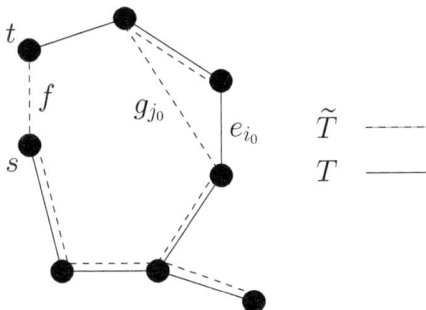

Fig. 3.3 Replacing f by e_{i_0} yields another optimal tree in the proof of Lemma 4

We complete the proof by showing that the cut criterion implies the circuit criterion. Since the circuit criterion implies optimality, sufficiency of the cut criterion follows. Thus let $f = (s, t) \in E \setminus T$ and $e_1, \ldots, e_k$ the s-t-path in T. Then for all i, $(T \setminus \{e_i\}) \cup \{f\}$ is a tree. By the cut criterion we have $w(e_i) \le w(f)$, implying the circuit criterion.   □

Now we give the reduction of Problem 1 to the case of non-negative weight functions.

Lemma 5 *Let $G_1 = (V, E, w)$ be an instance of Problems 1 and 2.*

(i) *If F is an optimal solution of Problem 1 such that $|F|$ is maximal, then F decomposes into $F = T \dot\cup N$ an optimal solution T of Problem 2, and non-positive edges $N = \{e \in E \setminus T \mid w(e) \le 0\}$.*

(ii) *If on the other hand T is an optimal solution of Problem 2 and $N = \{e \in E \setminus T \mid w(e) \le 0\}$, then $F = T \dot\cup N$ is an optimal solution of Problem 1 of maximum cardinality.*

Proof Let F be as in (i), and T a minimum spanning tree of the restricted instance $G_2 = (V, F, w)$ of the problem. We will show that T is an optimal solution of Problem 2 for the instance G_1. Assume to the contrary, that there is a tree $\widetilde{T} \subseteq E$ spanning G_1 and $w(\widetilde{T}) < w(T)$. Set

$$\widetilde{F} := (F \cup \widetilde{T}) \setminus (T \setminus \widetilde{T}) = (F \setminus T) \cup \widetilde{T}.$$

Then the graph $(V, \widetilde{F})$ is connected and

$$w(\widetilde{F}) = w(F) + w(\widetilde{T}) - w(T) < w(F)$$

contradicting optimality of F. Note, edges in $F \setminus T$ should have non-positive weights.

On the other hand let F, T and N be as in (ii). and assume there were a solution $\widetilde{F}$ of Problem 1 such that $w(\widetilde{F}) < w(F)$ or $w(\widetilde{F}) = w(F)$ and $|\widetilde{F}| > |F|$. Let $\widetilde{T} \subseteq \widetilde{F}$ be a minimum spanning tree of the graph restricted to $\widetilde{F}$ and chosen such that $|T \cap \widetilde{T}|$ is maximum. Then $w(T) \le w(\widetilde{T})$ and thus

$$\begin{aligned} w(T) + w(N) &= w(F) \\ &\ge w(\widetilde{F}) \\ &= w(\widetilde{T}) + w(\widetilde{F} \setminus \widetilde{T}) \\ &\ge w(T) + w(\widetilde{F} \setminus \widetilde{T}). \end{aligned}$$

This yields

$$0 \ge w(N) \ge w(\widetilde{F} \setminus \widetilde{T}).$$

On the other hand we have $|\widetilde{F} \setminus \widetilde{T}| \ge |N|$ and

$$\text{if } w(N) = w(\widetilde{F} \setminus \widetilde{T}), \text{ then} |\widetilde{F} \setminus \widetilde{T}| > |N|.$$

Hence $\widetilde{F} \setminus \widetilde{T}$ contains an edge f of non-positive weight that does not belong to N and hence $f \in T \setminus \widetilde{T}$. Choose such an $f = (s, t)$ such that $w(f)$ is minimum. The following part of the proof is similar to the proof of Lemma 4. Actually Fig. 3.3 applies with the roles of T and $\widetilde{T}$ interchanged. Now, applying the circuit criterion to $\widetilde{T}$ in $\widetilde{F}$ we conclude that the weight of f is at least the weight of any other edge on the s-t-path $e_1, \ldots, e_{k_f}$ in $\widetilde{T}$. As T is a tree, there is some $e_{i_0} = (u, v) \notin T$. Then we have $e_{i_0} \in \widetilde{T} \setminus T$ and $w(e_{i_0}) \le w(f)$. Let $g_1, \ldots, g_k$ be the unique (u, v)-path in T and $g_{j_0} \in T \setminus \widetilde{T}$. As T is optimal we conclude $w(g_{j_0}) \le w(e_{i_0}) \le w(f)$. By the choice of f we also have $w(f) \le w(g_{j_0})$ and thus $w(e_{i_0}) = w(f)$. Hence $\widehat{T} := \widetilde{T} \setminus \{e_{i_0}\} \cup \{f\}$ is a spanning tree of $\widetilde{F}$ satisfying $w(\widehat{T}) = w(\widetilde{T})$ and $|\widehat{T} \cap T| > |\widetilde{T} \cap T|$ contradicting the choice of $\widetilde{T}$. $\square$

3.4 Kruskal's Algorithm

After these preparations we arrive at the following algorithm to solve the minimal spanning tree problem.

ALGORITHM Kruskal

```
E = Sort(Edges, byWeight)

for e in E:
    if not CreatingCircuit(e):
5        AddEdge(e)
```

Theorem 1 *Kruskal's algorithm yields a minimum spanning tree T if run on a connected weighted graph $G = (V, E)$.*

Proof As G is connected and an edge e is not included in T only if there already exists a path between their endpoints, clearly T must be connected as well. Since T satisfies the circuit criterion the claim follows. $\square$

Let us analyze this algorithm. Sorting the edges requires $O(|E| \log |E|)$[1], the remaining loop is done in $O(|E| c(n))$ where $c(n)$ denotes the time to find out whether e_i closes a circuit with T. This can be done in a naive way in $O(|V|)$.

ALGORITHM KruskalFindCircuit

```
def CreatingCircuit(u,v):
2        path = ShortestPath(u,v)
    if path:
        return True
    else:
        return False
7
E = Sort(Edges, byWeight)

for (u,v) in E:
    if not CreatingCircuit(u,v):
12       AddEdge(u,v)
```

---

[1] Sorting algorithms are explained in most of the introductory algorithm textbooks. See for example [11].

There is a more efficient way to do that. Before discussing this in detail it is a good moment to have a break and do some computer experiments.

Software Exercise 16 Start CATBox and load `KruskalFindCircuit.alg` and `Kruskal1.cat`. This is already a tree and thus we will not find a circuit anywhere. We prepared a breakpoint on the statement **if not** `CreatingCircuit` `(u,v):`. When the algorithm stops there, you can push the trace button to step through the function `CreatingCircuit` that searches for a circuit; we call that *tracing a function*. We observe that in order to decide whether the new edge *e* closes a circuit we search for a path in the tree connecting its end vertices. This is done by a labeling procedure. It is easier to see with a graph like `Kruskal3.cat`. When edge $(4, 9)$ is checked we trace to see that a BFS is started at vertex 4 which terminates when 9 is labeled. This clearly takes $O(|V|)$.

Implementing the circuit test as described above amounts to computing the connected components for the tree edges from scratch for each tested edge. Once we know the components we can perform the circuit test in constant time by checking whether its end vertices belong to the same component. To save this effort we introduce a *data structure* which allows to perform the circuit test in constant time. We store the components—that is, the trees of the forest *F*—by associating with each vertex v a reference `component[v]` to a list of the vertices of the component and for each component a variable storing its size `Order(component[v])`. An edge $e_i = (u, v)$ closes a circuit if and only if `component[v]=component[u]` which can be checked in constant time. If e_i does not close a circuit we merge `component[v]` and `component[u]`. In order to do the merging efficiently we add the vertices of the smaller component to the larger one. We can find out which is smaller in constant time by comparing `Order(component[u])` to `Order` `(component[v])`.

Remark 3 The above implementation describes a simple *Union-Find* data structure [11]. The purpose of the Union-Find Problem is to organize the tasks of locating elements in a partition and merging classes of the partition efficiently.

Software Exercise 17 This method is visualized in `KruskalTrace.alg`. In the beginning all vertices have different colors[2]. The circuit test is very simple now, since we only have to compare two numbers. If an edge is added because it does not close a circuit, we have to work a bit harder. Then, the components are merged as described earlier.

---

[2] If the graph is of large order, several vertices will be assigned the same color. Picking a large set of colors which are pair-wise distinguishable is a non-trivial task.

ALGORITHM KruskalTrace

```
def AddEdge(u,v):
    C = component[u]
    D = component[v]
    if Order(C) < Order(D):
        for w in Vertices(C):
            component[w] = D
        MergeComponents(C,D)
    else:
        for w in Vertices(D):
            component[w] = C
        MergeComponents(D,C)

def CreatingCircuit(u,v):
    if SameComponent(u,v):
        return True
    else:
        return False

E = Sort(Edges, byWeight)

for (u,v) in E:
    if not CreatingCircuit(u,v):
        AddEdge(u,v)
```

Let us analyze this method. A single update may cost up to $O(|V|)$. However, this cannot happen too often because the size of the component will double in that case. Therefore we take a more global view combining the cost of merging over all iterations. We call this *amortized costs*.

Lemma 6 *The amortized costs of merging the components are* $O(|V| \log |V|)$.

Proof We proceed by induction on $n = |V|$. For $n = 1$ there is nothing to prove, thus we may assume that edge e merges the components T_1 and T_2 of sizes $n_1 \leq n_2$ where $n = n_1 + n_2$. Then $n_1 \leq \frac{n}{2}$ and the update is done in cn_1. This is added to the cost of merging operations to create T_1 and T_2, which by inductive assumption is $cn_1 \log_2 n_1$ respectively $cn_2 \log_2 n_2$. Summing this up yields

$$
cn_1 + cn_1 \log_2 n_1 + cn_2 \log_2 n_2 \leq cn_1 + cn_1 \log_2 \frac{n}{2} + cn_2 \log_2 n
$$
$$
= cn_1 + cn_1(\log_2 n - 1) + cn_2 \log_2 n
$$
$$
= cn \log_2 n.
$$

$\square$

3.5 Prim's Algorithm

Kruskal's algorithm decreases the number of components by using the cheapest edges linking disconnected subgraphs in each step. Its optimality is guaranteed by the circuit criterion. The approach to be discussed now grows a single non-trivial component. That this also yields a minimum spanning tree is guaranteed by the cut criterion. For each vertex v outside of, but reachable from the present tree T

with a single edge, we keep the cheapest such edge in a set F. In the beginning F equals the neighborhood of s. In an iteration of F we remove the cheapest edge from F, add it to T, which now contains an additional vertex v. For all neighbors w of v we check, whether $w \notin T$ and whether (v, w) has a weight smaller than the previous possibility e to connect T to w, where we follow the convention of using ∞ to denote the absence of such a possibility. If so, we exchange e for (v, w) and continue until T is spanning.

ALGORITHM Prim

```
   s = PickVertex()
 2 for v in Neighborhood(s):
       F.AddEdge((s,v))
       pred[v] = s

   while not T.IsSpanning():
 7     (u,v) = F.MinimumEdge(weight)
       F.DeleteEdge((u,v))
       T.AddEdge((u,v))

       for w in Neighborhood(v):
12         if weight[(pred[w],w)] > weight[(w,v)] \
               and not T.Contains(w):
               F.DeleteEdge((pred[w],w))
               F.AddEdge((w,v))
               pred[w] = v
```

Theorem 2 *Prim's algorithm yields a minimum spanning tree when run on a connected weighted graph.*

Proof In the routine T.AddEdge((v,w)) vertex w is added as a new vertex into the non-trivial component. In particular the **while**-loop is executed exactly $|V| - 1$ times. Thus the procedure is finite. Obviously, the subgraph $H = (V(T), E(T))$ is a tree throughout the runtime. We show by induction on $|T|$:

Any such tree T is contained in a minimum spanning tree of G.

The assertion is trivial if $T = (\{v\}, \emptyset)$. Thus let $|T| = k > 0$ and $e_0 = (s, u)$ be the last edge that has been added to T. By inductive assumption $T \setminus \{e_0\}$ is contained in a minimum spanning tree $\widetilde{T}$ of G. Let S denote the vertex set in the non-trivial component of $T \setminus \{e_0\}$ (respectively the vertex s if $T = \{e_0\}$). According to the selection rule in the algorithm all edges $(\widetilde{s}, l)$ such that $\widetilde{s} \in S$ and $l \in V \setminus S$ satisfy $w((\widetilde{s}, l)) \geq w(e_0)$. If $e_0 \in \widetilde{T}$, we are done. Otherwise let $g_1, \ldots, g_k$ denote the unique s-u-path in $\widetilde{T}$. Let g_{j_0} denote the first edge on this path visiting a vertex in $V \setminus S$. Then $w(g_{j_0}) \geq w(e_0)$ and thus $\widehat{T} := \widetilde{T} \setminus \{g_{j_0}\} \cup \{e_0\}$ is a tree spanning G and containing T. Furthermore it is minimum because $w(\widehat{T}) \leq w(\widetilde{T})$. $\square$

Analyzing the complexity of that procedure we account $O(|V|)$ for the first **for**-loop. In the **while**-loop we have to determine the minimum of a set of $O(|V|)$ in altogether $(|V| - 1)$ runs. In the inner **for**-loop we have to access all $O(|E|)$ edges in total once. If we keep F in a suitable data structure (say a heap, which you may recall from your introductory computer science class. Details can be found in [11]) we can determine the minimum and update the data structure in $O(\log(|V|))$. Altogether we arrive at a running time of $O((|E| + |V|) \log(|V|)) = O(|E| \log(|V|))$.

Exercise 18 For the efficiency of Kruskal's algorithm it is important that our implementation always updates the smaller component. This can be easily visualized with a modification of Kruskal's algorithm, which does not take component sizes into account when doing the merge. You can create such a variant by modifying `KruskalTrace.alg`. Create a copy of `KruskalTrace.alg` and call it `KruskalInefficient.alg`. Do the same for the prolog `KruskalTrace.pro`.

You now have to modify the function `AddEdge` in `KruskalInefficient.alg`. Remove, or comment out, the comparison of component orders and change the code to always merge components `C` and `D` and setting `component[w]` = `C` for all vertices in `D`. Do not forget that block structure is indicated by indentation in Python. The only change in `KruskalInefficient.pro` concerns the location of the default breakpoint. Now the statement **if not** `CreatingCircuit` `(u,v):` is found on line number 17. Change `breakpoints = [2, 22]` accordingly.

The difference between the two versions of the algorithm becomes very obvious when you run both of them on the graph `spiral.cat`.

3.6 Remarks

Kruskal's algorithm is also known as the *greedy algorithm*. This is better understood when considering the maximal spanning tree problem. Greedily, we always grab the heaviest feasible edge. Kruskal's article has been published in 1956 [29]. However, algorithms for solving the minimal spanning tree problem were already known to Borůvka (1926) [4] and Jarnik (1930) [25] from Czechia.

The cut criterion immediately implies the correctness of the following "dual" approach.

ALGORITHM MatroidDualKruskal

```
E = Sort(Edges, byDecreasingWeight)


5  for e in E:
       if not DisconnectingGraph(e):
           RemoveEdge(e)
```

We do not know about an efficient data structure to implement the test for connectivity here, that would make this approach competitive with Kruskal's algorithm.

When run on a disconnected weighted graph, Kruskal's algorithm yields a minimum spanning forest. To make Prim's algorithm work for the unconnected case as well we have to include edges with an infinitely large weight to make the graph connected. Deleting these artificial edges from the minimum spanning tree yields a minimum spanning forest.

The greedy algorithm does not only apply to minimal spanning trees. Consider the vertex-edge incidence matrix of a graph. Realize that the spanning trees correspond to exactly the bases of the matrix (i.e., the maximal sets of independent columns) over the binary field, usually called $GF(2)$ or $\mathbb{Z}_2$, the field with two elements where $1 + 1 = 0$. Here, the greedy algorithm yields a cheapest basis of a finite set of weighted vectors.

The abstract combinatorial structure, where the greedy algorithm always yields an optimal solution for any weight function is called a *matroid*. The following series of exercises develops the necessary background on matroids to repeat the proof of sufficiency of the circuit criterion in that setting. Once the proof is understood it will look simpler at that higher level of abstraction. As a consequence it immediately follows that the greedy algorithm computes a minimum cost basis of a matroid.

Exercise 19 A *matroid M* on a finite set E is an ordered pair $(E, \mathcal{C})$ where $\mathcal{C} \subseteq 2^E$ is a family of sets, called *circuits*, satisfying the following set of axioms.

C1) $\emptyset \notin \mathcal{C}$.
C2) $C_1 \in \mathcal{C}$ and $C_1 \subseteq C_2 \in \mathcal{C} \Rightarrow C_1 = C_2$.
C3) If $C_1, C_2 \in \mathcal{C}$ where $C_1 \neq C_2$ and $e \in C_1 \cap C_2 \Rightarrow \exists C_3 \in \mathcal{C} : C_3 \subseteq (C_1 \cup C_2) \backslash e$.

A set $I \in 2^E$ is called *independent*, if it does not contain a circuit, i.e. $C \not\subseteq I$ for all $C \in \mathcal{C}$ and is called *dependent* otherwise. Thus, the circuits are the minimal dependent sets. We denote the family of all independent sets by $\mathcal{I}$.

  (i) Let $G = (V, E)$ be a graph and $\mathcal{C}$ its family of simple circuits. Prove: $(E, \mathcal{C})$ is a matroid, the *graphic matroid M(G)*. What are the independent sets in that matroid?
 (ii) Let E denote the set of indices of a matrix $A \in \mathbb{R}^{m \times n}$ and $\mathcal{C}$ the set of all minimal subsets $C \subseteq E$, such that the set of columns indexed by C is linearly dependent in $\mathbb{R}^m$. Prove: $M = (E, \mathcal{C})$ is a matroid (called *vector matroid*, and denoted by $M[A]$).
(iii) Show that the graphic matroid $M[G]$ of a graph $G = (V, E)$ equals the vector matroid $M[A]$ of its incidence matrix considered as a matrix over $GF(2)$ the field with 2 elements.

Exercise 20 Let $M = (E, \mathcal{C})$ be a matroid and $w : E \rightarrow \mathbb{N}$ a non-negative weight function on the set of its elements. Prove that:

 (i) If $I \in \mathcal{I}, e \in E$ and $I \cup e \notin \mathcal{I}$, then there exists a unique circuit $C(I, e) \subseteq I \cup e$ and furthermore $e \in C(I, e)$ (called *fundamental circuit of e*).
(ii) The greedy algorithm computes an (inclusion-wise) maximal independent set of minimum weight.
     Hint: Modify the proof of the circuit criterion.

Exercise 21 Let $P = \{p_1, \ldots, p_k\} \subseteq \mathbb{R}^n$ be a finite set of vectors in Euclidean space and $w : P \rightarrow \mathbb{N}$ a weight function. Prove that a maximal linear independent set of maximum weight can be chosen greedily.

Exercise 22 Let $G = (V, E, w)$ be a weighted graph. Write an algorithm that computes a connected subgraph $H = (V, F)$, such that

$$\sum_{e \in F} w(e)$$

is maximal.

3.7 Some Additional Notation

In Proposition 3 we proceeded by induction and considered the graph that arose from the original one by removing a vertex. We will now introduce this concept and two other canonical subgraphs more formally.

Definition 7 Let $G = (V, E)$ be a graph, $W \subseteq V$ and $e = (u, v) \in E$.

(i) The graph $G[W] = (W, F)$ with edge set

$$F := \{f \in E \mid \varphi_G(f) \subseteq W\},$$

   i.e. edges both of whose end vertices belong to W, is called *subgraph induced by W*.

(ii) The graph $G \setminus \{e\} = (V, E \setminus e)$ arises from G by *deletion* of e.

(iii) Let $w \notin V$. The graph $G / e = (V \setminus \{u, v\} \cup \{w\}, E \setminus e)$ with incidence function

$$\varphi_{G/e}(f) = \begin{cases} \varphi(f) & \text{if } \varphi(f) \cap \{u, v\} = \emptyset \\ \varphi(f) \setminus \{u, v\} \cup \{w\} & \text{else} \end{cases}$$

   arises by *contraction* of e (see Fig. 3.4).

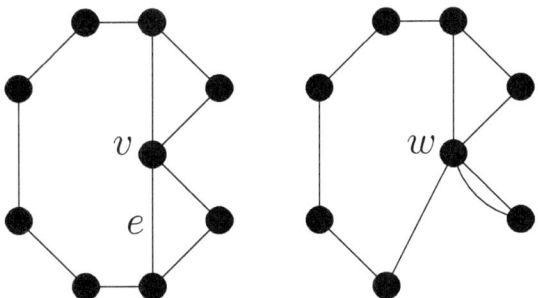

Fig. 3.4 A graph G and its contraction G / e

Exits

Surprisingly enough, although the problem appears to be quite simple, algorithms for minimum spanning trees are still an active field of research. In particular, it is an open problem whether there exists an algorithm with a linear running time bound. Presently, the best bound achieved is $O(\alpha(m)n)$ where α is the inverse of the so-called Ackermann function, a very fast growing function such that its inverse is "almost" a constant. For a recent survey on the history and the future of the MST-problem we refer the reader to [33].

Exercises

Exercise 23 Contraction and deletion are commutative and associative. If $F_1, F_2 \subseteq E$ are disjoint edge sets then $G / F_1 \setminus F_2$ is well defined.

Exercise 24 Let $M = (E, \mathcal{C})$ be a matroid and $e \in E$. Prove that

(i) $M \setminus e := (E \setminus e, \mathcal{C} \setminus e)$, where $\mathcal{C} \setminus e := \{C \in \mathcal{C} : e \notin C\}$, is a matroid.
(ii) $M / e := (E \setminus e, \mathcal{C} / e)$ such that $\mathcal{C} / e := \text{elem}(\{C \setminus e : C \in \mathcal{C}\})$ is a matroid, where $\text{elem}(\mathcal{F}) := \{\emptyset \neq S \in \mathcal{F} : \nexists T \in \mathcal{F} \text{ with } T \subsetneq S\}$ for any family $\mathcal{F}$ of sets.

As in the case of graphs, these objects are called *minors* of M. They arise by *contraction* respectively *deletion* of e.

Chapter 4
Linear Programming Duality

Some algorithms which we shall introduce are called primal-dual algorithms. This concept cannot be fully understood without the notion of linear programming duality. The reader not interested in the more advanced algorithms of this book can feel free to skip this chapter. The mathematics of this chapter is inevitable though, to fully appreciate the algorithms for mincost-flow and weighted matching.

4.1 The MST-Polytope

Already the greedy algorithm (Kruskal) to compute minimum spanning trees can be considered in the framework of primal-dual algorithms. Since the algorithm itself is quite simple, we will use it as our guiding example to introduce the theory of primal-dual algorithms. First we turn our combinatorial problem into a geometric one. For a set E and a subset $T \subseteq E$ the characteristic function χ_T indicates whether its argument is an element of T, more formally:

Definition 8 Let E be a set and $T \subseteq E$. We define the *characteristic function* $\chi_T :$ $E \to \{0, 1\}$ as

$$\chi_T(e) = \begin{cases} 1 \text{ if } e \in T \\ 0 \text{ if } e \notin T \end{cases} .$$

If $|E|$ is finite, we frequently call χ_T an *incidence vector*.

Example 2 Consider the graph in Fig. 4.1 and therein the spanning tree that is depicted by the dashed lines. The incidence vector of the set of edges of this spanning tree is given by $(1, 1, 1, 1, 1, 0, 0, 1, 1, 1)$.

This way we can consider any spanning tree of a fixed graph G as a point in $\mathbb{R}^{|E|}$. Therefore we can state the task of finding a minimum spanning tree in the following way:

Problem 3 Determine a tree T, such that χ_T minimizes the linear function $w^\top \chi_T :=$ $\sum_{e \in E} w_e \chi_T(e)$.

W. Hochstättler, A. Schliep, *CATBox*, DOI 10.1007/978-3-642-03822-8_4,
© Springer-Verlag Berlin Heidelberg 2010

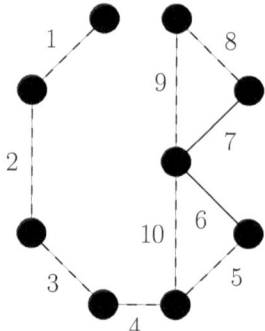

Fig. 4.1 Example 2

Up to now we did nothing but reformulate the problem. In the sequel, we will modify the problem, because it is relatively easy to optimize a linear function over a *polyhedron*—which we will define in Definition 10. To exploit this fact, we have to turn the discrete problem into a continuous one.

Definition 9 (Hull operators from linear algebra) Let $x_1, \ldots, x_k \in \mathbb{R}^n$ and $\lambda \in \mathbb{R}^k$. Then

$$x = \sum_{i=1}^{k} \lambda_i x_i$$

is called a *linear combination* of $x_1, \ldots, x_k$. If, in addition, we have

$$\left\{ \begin{array}{l} \lambda \geq 0 \\ \sum_{i=1}^{k} \lambda_i = 1 \\ \lambda \geq 0, \ \sum_{i=1}^{k} \lambda_i = 1 \end{array} \right\}, \text{ then x is called a} \left\{ \begin{array}{l} \text{conic} \\ \text{affine} \\ \text{convex} \end{array} \right\} \text{ combination.}$$

The notation $\lambda \geq 0$ means that λ is non-negative in each coordinate, i.e. $\lambda_i \geq 0$ for $i = 1, \ldots, k$. For $S \subseteq \mathbb{R}^n$ we denote by

$$\left\{ \begin{array}{l} \text{lin} \,(S) \\ \text{cone}\,(S) \\ \text{aff}\,(S) \\ \text{conv}\,(S) \end{array} \right\}, \text{ the set of all} \left\{ \begin{array}{l} \text{linear} \\ \text{conic} \\ \text{affine} \\ \text{convex} \end{array} \right\} \text{ combinations of}$$

elements from S. Note that combinations are only defined for finite subsets of S. We call these sets hulls (cf. Fig. 4.2).

Now we can consider the convex hull of the incidence vectors of the spanning trees of a graph G,

$$P(G) := \text{conv}\,\{\chi_T \in \mathbb{R}^E \mid T \text{ is a tree spanning } G\}.$$

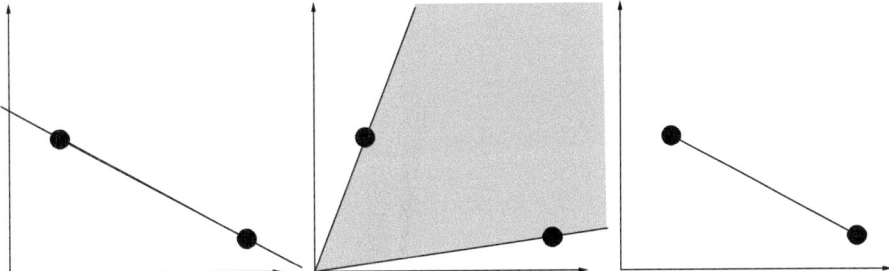

Fig. 4.2 Affine, conic and convex hull of two points in the plane

To be able to optimize over this object with standard methods from linear optimization we need a different description of $P(G)$, namely as an intersection of *affine
half spaces*. An infinite line, given by an equation $a_1 y + a_2 x = b$ divides the plane
into two halves, $a_1 y + a_2 x < b$ respectively $a_1 y + a_2 x > b$. Each half, together with
the line, forms an affine half space in the plane. In higher dimension $\mathbb{R}^n$ the same
applies to affine hyperplanes given by an equation $a^\top x := a_1 x_1 + \ldots + a_n x_n = b$,
where $a = (a_1, \ldots, a_n)$ and $x = (x_1, \ldots, x_n)$ are vectors. An intersection of finitely
many affine half spaces is called a polyhedron (see Fig. 4.4):

Definition 10 A set $P \subseteq \mathbb{R}^n$ is called a *polyhedron*, if there exists an $m \in \mathbb{N}$, an
$\mathbb{R}^{m \times n}$ matrix A and some vector $b \in \mathbb{R}^m$ such that

$$P = \{x \in \mathbb{R}^n \mid Ax \leq b\}.$$

To obtain a representation of our collection X of the spanning trees, as a polyhedron $P(G)$ we consider the following set. Let $G = (V, E)$ be a connected graph,
$\mathcal{P}$ the set of all partitions of V. Recall that a *partition* $V = V_1 \dot{\cup} \ldots \dot{\cup} V_k$ of a set V
is a decomposition of the set into non-empty disjoint subsets. Each of these subsets
is called a *class*. For $P \in \mathcal{P}$ let ∂P be the set of all edges that connect vertices
from different classes of the partition and denote by $|P|$ the number of classes in P.
Given a graph G we can define (see Fig. 4.3)

$$ST(G) := \{x \in \mathbb{R}^E \mid \forall P \in \mathcal{P} : \sum_{e \in \partial P} x_e \geq |P| - 1 \text{ and } \sum_{e \in E} x_e = |V| - 1, x \geq 0\}.$$

The characteristic vector of any spanning tree will satisfy these relations because,
when we identify the classes of the partition with new (super)-nodes, the edges
connecting them will still form a connected graph. As we have $|P|$ super nodes any
such graph has to have at least $|P| - 1$ vertices. This implies that $P(G) \subseteq ST(G)$ as
we will show later. Moreover we will show that actually the sets $ST(G)$ and $P(G)$
are equal.

Polyhedra are *convex* objects, meaning that any convex combination of a set of
points from a polyhedron is a point of the polyhedron again.

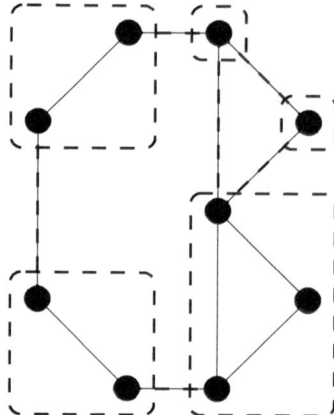

Fig. 4.3 A partition P and ∂P. The *rectangles* represent individual classes of the partition. *Edges* incident with vertices from the same class are drawn solid, those from distinct classes *dashed*

Lemma 7 *Let P be a polyhedron $v_1, \ldots, v_k \in P$ and $\lambda \in \mathbb{R}^k$ such that $\lambda \geq 0$ and $\sum_{i=1}^{k} \lambda_i = 1$. Then*

$$\sum_{i=0}^{k} \lambda_i \, v_i \in P.$$

Proof Let $A \in \mathbb{R}^{m \times n}, b \in \mathbb{R}^m$ such that $P = \{x \in \mathbb{R}^n \mid Ax \leq b\}$. Then

$$\forall \, i = 1, \ldots, k : Av_i \leq b \Rightarrow \forall \, i = 1, \ldots, k : A \lambda_i v_i \leq \lambda_i b$$

$$\Rightarrow A \left(\sum_{i=1}^{k} \lambda_i v_i \right) = \sum_{i=1}^{k} A \lambda_i v_i \leq \sum_{i=1}^{k} \lambda_i b = b.$$

$\square$

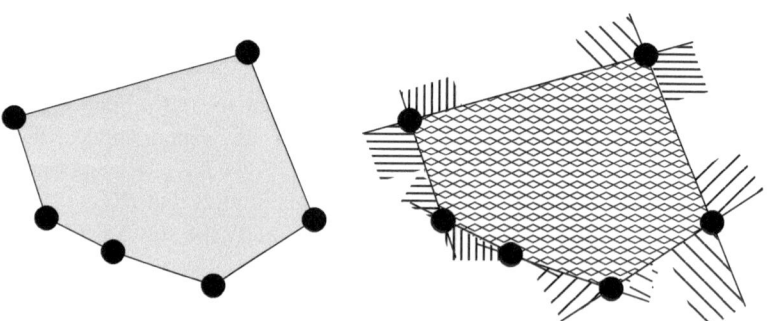

Fig. 4.4 A polyhedron as convex hull of points and as intersection of half spaces

Now we start to show that the convex hull of the incidence vectors of spanning trees forms a polyhedron. More precisely,

Theorem 3

$$P(G) = ST(G).$$

Proof "$\subseteq$" Let T be a spanning tree, χ_T its incidence vector and $P \in \mathcal{P}$ a partition. If we contract all the vertices and edges in each class to a single node, then the tree edges connecting the classes must form a connected graph. In particular this graph must still contain a tree, and thus have at least $|P| - 1$ edges and $\sum_{e \in \partial P}(\chi_T)_e \geq |P| - 1$ is clearly satisfied. Since any tree has $|V| - 1$ edges and, obviously, $\chi_T \geq 0$ the claim follows from Lemma 7.

"$\supseteq$": To prove the other inclusion we have to work a bit harder. Let $x \in ST(G)$. First we show that $x \leq 1$. Suppose to the contrary that $x_e > 1$ for some edge $e = (u, v)$. Consider the partition P_e for which $\{u, v\}$ forms one class and all other classes are *trivial*; that is, they consist of single vertices. Then $e \notin \partial P$, and thus

$$\sum_{f \in \partial P_e} x_f \leq |V| - 1 - x_e < |V| - 2 = |P_e| - 1,$$

a contradiction, since $\sum_{e \in \partial P_e} x_e \geq |P_e| - 1$ for *all* possible partitions.

In the following we proceed by induction on the number of nonzero entries of x. The set of indices of those nonzero entries is called the *support* of x, denoted $\text{supp}(x)$. As each entry of x is at most one, x must contain at least $|V| - 1$ such non-zero entries: $|\text{supp}(x)| \geq |V| - 1$. Let $G(x) = (V, \text{supp}(x))$ denote the subgraph of G consisting of the edges that have a non-zero entry in x. Note that $G(x)$ is connected. Otherwise, the components of $G(x)$ define classes for a non-trivial partition P of V, which violates—as $\sum_{e \in \partial P} x_e = 0 < |P| - 1$—the assumption $x \in ST(G)$. In particular, if $G(x)$ contains exactly $|V| - 1$ edges it is itself a tree and hence $x \in P(G)$ which founds the induction.

We treat the case that x is on the boundary of $ST(G)$—i.e., some inequality is not strict—separately from the case that x falls into its interior. We start with the former.

Lemma 8 *Assume Theorem 3 is true for all graphs with less than $|V|$ vertices, $x \in ST(G)$ and there is a non-trivial partition P of the vertex set V such that*

$$\sum_{e \in \partial P} x_e = |P| - 1. \tag{4.1}$$

Then $x \in P(G)$.

Proof Among the partitions for which x satisfies (4.1) we choose a $P \in \mathcal{P}$ that has as few non-trivial classes as possible and show:

P contains exactly one non-trivial class.

Assume that the classes of P are sorted by size in decreasing order, that is $P = V_1 \dot\cup \ldots \dot\cup V_{|P|}$ and $|V_1| \geq \ldots \geq |V_k| \geq 2$ while $|V_{k+1}| = \ldots = |V_{|P|}| = 1$. As the classes $V_{k+1}, \ldots, V_{|P|}$ are trivial we only have edges in ∂P, for which x_e sum to $|P| - 1$, and in the non-trivial classes $V_1, \ldots, V_k$. Hence, we can split the sum $\sum_e x_e$ and obtain

$$\sum_{i=1}^{k} \sum_{e \subseteq V_i} x_e = |V| - 1 - (|P| - 1).$$

Employing an identity which generally holds for partitions,

$$|V| - l = \sum_{j=1}^{l} (|Q_i| - 1)$$

where $V = Q_1 \dot\cup \ldots \dot\cup Q_l$ is an arbitrary partition, we finally derive

$$\sum_{i=1}^{k} \sum_{e \subseteq V_i} x_e = |V| - |P| = \sum_{i=1}^{|P|} (|V_i| - 1) = \sum_{i=1}^{k} (|V_i| - 1). \qquad (4.2)$$

If there exists an index $j \in \{1, \ldots, k\}$ such that $\sum_{e \subseteq V_j} x_e > |V_j| - 1$, then we consider another partition P_j, which consists of the non-trivial class V_j and a trivial class for each of the remaining vertices. The partition P_j satisfies

$$\sum_{e \in \partial P_j} x_e = |V| - 1 - \sum_{e \subseteq V_j} x_e < |V| - 1 - (|V_j| - 1) = |P_j| - 1,$$

a contradiction. Therefore (4.2) implies that we must have

$$\sum_{e \subseteq V_j} x_e = |V_j| - 1 \qquad (4.3)$$

in each class. In particular we have

$$\sum_{e \in \partial P_1} x_e = |V| - 1 - (|V_1| - 1) = |P_1| - 1,$$

implying $k = 1$ by the choice of P. Thus, P is in fact equal to P_1.

Now, let $G[V_1] = (V_1, F)$ denote the graph *induced on* V_1, i.e. the graph with vertex set V_1 and edge set $F = \{(u, v) \in E \mid \{u, v\} \subseteq V_1\}$, and decompose $x = x_1 + x_2$ into two vectors, where $\mathrm{supp}(x_1) \subseteq F$ and $\mathrm{supp}(x_2) \cap F = \emptyset$. This means, we decompose the entries of x into those of edges inside and those of edges outside V_1. Let $x_1' \in \mathbb{R}^F$ denote the restriction of x_1 to the coordinates of F and x_2' the restriction of x_2 to $\mathbb{R}^{E \setminus F}$.

We claim that $x'_1 \in ST(G[V_1])$. Let R be a partition of V_1. We extend this to a partition $\widetilde{P}$ of V by adding the vertices from $V \setminus V_1$ as trivial classes. As $x \in ST(G)$ it must satisfy

$$\sum_{e \in \partial \widetilde{P}} x_e \geq |\widetilde{P}| - 1. \tag{4.4}$$

Now using $\sum_{e \in \partial P_1}(x_2)_e = \sum_{e \in \partial P_1} x_e = |P_1| - 1$ we compute

$$
\sum_{e \in \partial R}(x'_1)_e = \sum_{e \in \partial R}(x_1)_e + \underbrace{\sum_{e \in \partial P_1}(x_2)_e - (|P_1| - 1)}_{=0}
$$

$$
= \sum_{e \in \partial \widetilde{P}} x_e - (|P_1| - 1)
$$

$$
\overset{(4.4)}{\geq} |\widetilde{P}| - 1 - (|P_1| - 1)
$$

$$
= |R| - 1.
$$

By inductive assumption we, thus, have trees $S_1, \ldots, S_s$ spanning V_1 and convex coefficients $\mu_1, \ldots, \mu_s \geq 0$ such that $\sum_{i=1}^{s} \mu_i = 1$ and $x'_1 = \sum_{i=1}^{s} \mu_i \chi_{S_i}$.

For x'_2 we consider the graph G/F that arises if we contract V_1 to a single vertex. A computation analogous to the above verifies $x'_2 \in ST(G/F)$. Hence by induction we find trees $U_1, \ldots, U_l$ spanning $ST(G/F)$, and coefficients such that

$$x'_2 = \sum_{j=1}^{l} \nu_j \chi_{U_j}$$

is a convex combination of their characteristic vectors. Now, setting $T_{i,j} := S_i \cup U_j$ for $i = 1, \ldots, s$ and $j = 1, \ldots, l$ we get trees spanning G and, setting $\lambda_{i,j} := \mu_i \nu_j$, we compute $\sum_{i,j} \lambda_{i,j} = \sum_{j=1}^{l} \nu_j = 1$ and if now we consider all vectors in $\mathbb{R}^E$ (i.e., extended by zeroes if necessary,)

$$
\sum_{\substack{i=1 \\ j=1}}^{k,l} \lambda_{i,j} \chi_{T_{i,j}} = \sum_{\substack{i=1 \\ j=1}}^{k,l} \lambda_{i,j}(\chi_{S_i} + \chi_{U_j})
$$

$$
= \sum_{\substack{i=1 \\ j=1}}^{k,l} \mu_i \nu_j \chi_{S_i} + \sum_{\substack{i=1 \\ j=1}}^{k,l} \mu_i \nu_j \chi_{U_j}
$$

$$= \sum_{i=1}^{k} \mu_i \chi_{S_i} + \sum_{j=1}^{l} \nu_j \chi_{U_j}$$

$$= x_1 + x_2 = x.$$

$\square$

Continuing with the proof of Theorem 3 we are left with the case that x is a point in the interior of our polyhedron. First, we choose an arbitrary tree T_0 spanning $G(x)$. Since $ST(G)$ is bounded, the line drawn through the characteristic vector χ_{T_0} of T_0 and x will eventually hit the boundary of $ST(G)$ in a second point y. This point y is a convex combination of trees spanning G by Lemma 8 or inductive assumption and x is a convex combination of χ_{T_0} and y. The following formalizes this geometric idea.

By assumption $\sum_{e \in \partial P} x_e > |P| - 1$ for all non-trivial partitions of V. Let T_0 be some tree spanning $G(x)$. Clearly, $x \neq \chi_{T_0}$ so there is an edge $e \in T_0$ such that $0 < x_e < (\chi_{T_0})_e = 1$. Consider the ray

$$y(\lambda) := \frac{1}{1 - \lambda} x - \frac{\lambda}{1 - \lambda} \chi_{T_0} \text{ for } \lambda \in [0, 1).$$

For $\lambda = 0$ we get x, and thus a point in $ST(G)$. For $\lambda = x_e$ we compute $y(x_e)_e = \frac{x_e}{1 - x_e} - \frac{x_e}{1 - x_e} = 0$, and $y(\lambda) \notin ST(G)$ for $x_e < \lambda < 1$. Besides this non-negativity constraint there may be another inequality, possibly coming from some partition Q, which might be violated. Say, such an inequality is violated by $y(x_e)$. There must however exist some maximal $\lambda_0 \in [0, x_e]$ such that $y(\lambda_0) \geq 0$ and

$$\sum_{f \in \delta P} y(\lambda_0)_f \geq |P| - 1 \qquad (4.5)$$

for all non-trivial partitions P of V. Furthermore,

$$\sum_{f \in E} y(\lambda_0)_f = \frac{1}{1 - \lambda_0}(|V| - 1) - \frac{\lambda_0}{1 - \lambda_0}(|V| - 1) = |V| - 1,$$

thus $y(\lambda_0) \in ST(G)$ and as λ_0 was chosen maximal, there is some inequality defining $ST(G)$ that is no longer strict. Hence, we either must have $y(\lambda_0)_e = 0$ for some $e \in T_0 \subseteq supp(x)$ or there exists $P \in \mathcal{P}$ such that $\sum_{e \in \partial P} y(\lambda_0)_e = |P| - 1$. In the first case $G(y(\lambda_0))$ has at least one edge less than $G(x)$ and $y(\lambda_0)$ is convex combination of spanning trees by inductive assumption, otherwise this holds by Lemma 8. In both cases we find $\mu_1, \ldots, \mu_k$ and spanning trees $T_1, \ldots, T_k$ such that $\tilde{x} := y(\lambda_0) = \sum_{i=1}^{k} \mu_i \chi_{T_i}$. Setting $\lambda_i = (1 - \lambda_0)\mu_i$, we get $\sum_{i=0}^{k} \lambda_i = \lambda_0 + (1 - \lambda_0)\sum_{i=1}^{k} \mu_i = 1$ and

$$\sum_{i=0}^{k} \lambda_i \chi_{T_i} = \lambda_0 \chi_{T_0} + (1 - \lambda_0) \sum_{i=1}^{k} \mu_i \chi_{T_i}$$
$$= \lambda_0 \chi_{T_0} + (1 - \lambda_0) y(\lambda_0)$$
$$= \lambda_0 \chi_{T_0} + x - \lambda_0 \chi_{T_0}$$
$$= x.$$

$\square$

Thus, we have turned the discrete Problem 2 into a linear problem of the following form:

$$\min w^\top x$$
$$\text{subject to } \sum_{e \in \partial P} x_e \geq |P| - 1 \; \forall P \in \mathcal{P}$$
$$\sum_{e \in E} x_e = |V| - 1$$
$$x \geq 0.$$

Generally, we call the task

$$\min c^\top x$$
$$\text{subject to } Ax \geq b$$
$$x \geq 0,$$

where $A \in \mathbb{R}^{m \times n}, b \in \mathbb{R}^m$ a *Linear Program* or an *LP*. There exists a formalism that, given a bounded linear program, yields a so called dual program, that has the same optimal value. We will present some theory to make this more precise in the next section.

4.2 Farkas' Lemma

Recall from Linear Algebra the definition of the orthogonal complement of a real vector space $L \subseteq \mathbb{R}^n$:

$$L^\perp := \{y \in \mathbb{R}^n \mid x^\top y = 0 \text{ for all } x \in L\}.$$

Lemma 9 (Farkas' Lemma) *Let $L \subseteq \mathbb{R}^n$ be a real vector space. Exactly one of the following alternatives holds.*

(i) $\exists x \in L : x_1 > 0, x \geq 0$
(ii) $\exists y \in L^\perp : y_1 > 0, y \geq 0$.

Proof Obviously, it is impossible for both alternatives to hold at the same time, for otherwise we had x, y such that $0 = x^\top y = \sum_{i=1}^{n} x_i y_i \geq x_1 y_1 > 0$.
    We will prove the remaining assertion by induction on n.

$n = 1$: Here either $L = \mathbb{R}$ or $L^\perp = \mathbb{R}$ and the assertion holds.

$n \geq 2$: Given $L \subset \mathbb{R}$, we consider the vector spaces in $\mathbb{R}^{n-1}$

$$\widetilde{L} := \{\tilde{x} \in \mathbb{R}^{n-1} \mid \exists x \in L, x_n = 0, x_i = \tilde{x}_i, i = 1, \ldots, n-1\}$$
$$\widehat{L} := \{\hat{x} \in \mathbb{R}^{n-1} \mid \exists x \in L, x_i = \hat{x}_i, i = 1, \ldots, n-1\}$$
$$\widetilde{L^\perp} := \{\tilde{y} \in \mathbb{R}^{n-1} \mid \exists y \in L^\perp, y_n = 0, y_i = \tilde{y}_i, i = 1, \ldots, n-1\}$$
$$\widehat{L^\perp} := \{\hat{y} \in \mathbb{R}^{n-1} \mid \exists y \in L^\perp, y_i = \hat{y}_i, i = 1, \ldots, n-1\}$$

Intuitively, $\widetilde{L}$ is the intersection of L with the coordinate hyperplane $x_n = 0$, $\widehat{L}$ the orthogonal projection onto the coordinate hyperplane and $\widetilde{L^\perp}$ and $\widehat{L^\perp}$ are their respective duals. First we show that $(\widetilde{L})^\perp = \widehat{L^\perp}$ and $(\widehat{L})^\perp = \widetilde{L^\perp}$. By symmetry it suffices to verify the former. Clearly, $\widehat{L^\perp} \subseteq (\widetilde{L})^\perp$. Let $\hat{y} \in (\widetilde{L})^\perp$. We have to show that there is some $y \in L^\perp$ which agrees with $\hat{y}$ in the first $n-1$ places. If $\sum_{i=1}^{n-1} x_i \hat{y}_i = 0$ for all $x \in L$, then $(\hat{y}, 0) \in L^\perp$ and we are done. Thus we may assume that there is some $x \in L$ such that $\sum_{i=1}^{n-1} x_i \hat{y}_i = c \neq 0$. Necessarily, then $x_n \neq 0$ and we claim that $y := (\hat{y}, \frac{-c}{x_n}) \in L^\perp$. Suppose to the contrary that $z \in L$ and $z^\top y \neq 0$ and thus $z_n \neq 0$. We may assume that z is scaled such that $(z - x)_n = 0$. Hence $\widetilde{(z - x)} \in \widetilde{L}$ and $y^\top(z - x) = y^\top z \neq 0$, a contradiction since $\hat{y} \in (\widetilde{L})^\perp$. Thus $\widehat{L^\perp} \supseteq (\widetilde{L})^\perp$.

Now we can proceed with the proof of Farkas' Lemma. We apply the inductive assumption for $\widetilde{L}$ and $\widehat{L}$. If there exists some $\tilde{x} \in \widetilde{L}$ such that $\tilde{x}_1 > 0, \tilde{x} \geq 0$, the corresponding x is as required. The same holds, if we find $\tilde{y} \in \widetilde{L^\perp}$ satisfying $y_1 > 0, y \geq 0$. Thus we may assume that in both cases, when we apply the inductive assumption, the other alternative holds; i.e., there exist $x \in L, y \in L^\perp$ such that $x_1, y_1 > 0$ and $x_i, y_i \geq 0$ for $i = 1, \ldots, n-1$. From $x^\top y = 0$ we derive

$$-x_n y_n = x_1 y_1 + \sum_{i=2}^{n-1} x_i y_i$$
$$\geq x_1 y_1$$
$$> 0.$$

This implies that exactly one of x_n and y_n must be strictly positive. If $x_n > 0$, then x is as required for the first alternative otherwise y for the second. $\qquad\square$

Remark 4 The same proof shows that the above theorem still holds true, if L and $L^\perp$ are subspaces of $\mathbb{Q}^n$.

Farkas' Lemma has different equivalent formulations. The most popular version probably is

If $P \subseteq \mathbb{R}^n$ is a polytope and $x \in \mathbb{R}^n$, then either $x \in P$ or there exists an affine hyperplane separating x and P.

Before putting this version of Farkas' Lemma into our algebraic notation we recall some facts from linear algebra.

Proposition 4 *Let $A \in \mathbb{R}^{m \times n}$. Then*

$$L = ker(A) := \{x \in \mathbb{R}^n \mid Ax = 0\}$$

and

$$im(A^\top) := \{x \in \mathbb{R}^n \mid \exists y \in \mathbb{R}^m : x = A^\top y\}$$

form an orthocomplementary pair of real vector spaces in $\mathbb{R}^n$; that is, $im(A^\top) = L^\perp$.

Now we formulate our separation theorem.

Corollary 1 *Let $v_1, \ldots, v_k, x \in \mathbb{R}^n$. Then:*
Either $x \in conv(\{v_1, \ldots, v_k\})$ or there exists $a \in \mathbb{R}^n, \beta \in \mathbb{R}$, such that $a^\top v_i \le \beta$ for $i = 1, \ldots, k$ and $a^\top x > \beta$.

Proof Let

$$A = \begin{pmatrix} -x & v_1 & \ldots & v_k \\ -1 & 1 & \ldots & 1 \end{pmatrix}$$

and apply Farkas' Lemma to the pair $L = ker(A)$, $L^\perp = im(A^\top)$. Then either there exists $\binom{\mu_0}{\mu} \in \mathbb{R}^{k+1}$, $\mu \ge 0$, $\mu_0 > 0$ such that $A\binom{\mu_0}{\mu} = 0$ or there exists $\binom{a}{a_{n+1}} \in \mathbb{R}^{n+1}$ such that $A^\top\binom{a}{a_{n+1}} \ge 0$ and $\left(A^\top\binom{a}{a_{n+1}}\right)_1 > 0$. The latter means $(-a)^\top x > a_{n+1}$ and $a^\top v_i \ge -a_{n+1}$, and thus $(-a)^\top v_i \le a_{n+1}$ for $i = 1, \ldots, k$. In the former case we put $\lambda_i := \frac{\mu_i}{\mu_0}$ resulting in $-x + \sum_{i=1}^{k} \lambda_i v_i = 0$ and $-1 + \sum_{i=1}^{k} \lambda_i = 0$, which verifies that $x \in conv(\{v_1, \ldots, v_k\})$. $\qquad\square$

Remark 5 If all data $v_1, \ldots, v_n, x$ in Corollary 1 is rational, then a, β can be chosen to be rational as well. This is an immediate consequence of Remark 4. The dividing hyperplane obtained is given in rational coordinates, which is not guaranteed by the more often presented proof using the derivative.

Exercise 25 Prove the following variants of Farkas' Lemma for $A \in \mathbb{R}^{m \times n}, b \in \mathbb{R}^m$, respectively suitable matrices and vectors B, C, D, a:

 (i) Either there exists a non-negative $x \in \mathbb{R}^n$ such that $Ax = b$ or there exists $u \in \mathbb{R}^m$ such that $u^\top A \le 0, u^\top b > 0$, but not both.
 (ii) Either there exists a non-negative $x \in \mathbb{R}^n$ such that $Ax \le b$ or there exists a non-negative $u \in \mathbb{R}^n$ such that $u^\top A \ge 0, u^\top b < 0$, but not both.
 (iii) Either there exists $x \in \mathbb{R}^n$ such that $Ax \le b$ or there exist a non-negative $u \in \mathbb{R}^m$ such that $u^\top A = 0, u^\top b < 0$, but not both.
 (iv) Either there exists $x \in \mathbb{R}^n$ such that $Ax = b$ or there exists $u \in \mathbb{R}^m$ such that $u^\top A = 0, u^\top b > 0$ but not both.
 (v) Either there exist x, y such that

$$Ax + By \leq a$$
$$Cx + Dy = b \qquad \text{or there exist } u, v \text{ such that}$$
$$x \geq 0$$

$$u^\top A + v^\top C \geq 0$$
$$u^\top B + v^\top D = 0$$
$$u^\top a + b^\top v < 0$$
$$u \geq 0,$$

but not both.

4.3 Duality Theorem of Linear Programming

Farkas' Lemma is at the heart of linear programming duality which we now can easily derive. Farkas' Lemma yields a "good characterization" of linear programming in the following sense. Consider the statement:

The linear program

$$(P) \qquad \begin{array}{c} \min c^\top x \\ \text{subject } Ax \geq b \\ \text{to} \quad x \geq 0 \end{array} \qquad (4.6)$$

has an optimal value of at most z_0.

The correctness of this statement is easily verified by presenting some $x \geq 0$ such that $Ax \geq b$ and $c^\top x \leq z_0$. If there is no such x, Farkas' Lemma serves as a tool to falsify the assertion. Namely, in this case the kernel of the matrix

$$B := \begin{pmatrix} -b & A & -I_m & 0 \\ -z_0 & c^\top & 0^\top & 1 \end{pmatrix}$$

cannot contain an element that is nonnegative in any component and strictly positive in the first one. For assuming (x_0, x, y, x_{n+1}) were such an element, we get $A\frac{x}{x_0} - \frac{y}{x_0} = b$, $c^\top \frac{x}{x_0} \leq z_0$. Therefore, by Farkas' Lemma, there exists a vector (v, t) such that $v^\top b + t z_0 < 0$, $v^\top A + t c^\top \geq 0$, $v \leq 0$, $t \geq 0$. Letting $u = -v$ we get

$$u^\top b > t z_0, \ u^\top A \leq t c^\top, \ u \geq 0, \ t \geq 0.$$

This motivates the following definition:

Definition 11 The linear program

$$(D) \qquad \begin{array}{c} \max u^\top b \\ \text{subject } u^\top A \leq c^\top \\ \text{to} \quad u \geq 0 \end{array}$$

is called the *dual program* to the *primal* linear program (P).

It is easily seen that the dual program bounds the primal from below:

Lemma 10 (Weak Duality) *If $x \geq 0$ is feasible for the primal program and u for the dual program then $c^\top x \geq u^\top b$.*

Proof

$$c^\top x \geq u^\top A x \geq u^\top b.$$

The first inequality holds since $c^\top \geq u^\top A$ and $x \geq 0$, the second since $Ax \geq b$ and $u \geq 0$. □

Theorem 4 (Strong Duality) *Let $A \in \mathbb{R}^{m \times n}, b \in \mathbb{R}^m$ and $c \in \mathbb{R}^n$. Then exactly one of the following four alternatives holds:*

 (i) *The primal program is feasible and bounded. In this case the dual program is feasible and bounded, too, and if z_0 denotes the supremum of the objective function of the primal program then there exists an optimal solution u^* of the dual such that $z_0 = u^{*\top} b$.*
 (ii) *The primal program is unbounded, i.e. there exists $x_0 \geq 0$ such that $Ax_0 \geq b$ and $x_1 \geq 0$ such that $Ax_1 \geq 0$ and $c^\top x_1 < 0$.*
 (iii) *The dual program is unbounded, i.e. there exists $u_0 \geq 0$ such that $u_0^\top A \leq c^\top, u_0 \geq 0$ and $u_1 \geq 0$ such that $u_1^\top A \leq 0$ and $u_1^\top b > 0$.*
 (iv) *Both programs are infeasible, i.e. there exists neither $x \geq 0$ such that $Ax \geq b$ nor $u \geq 0$ such that $u^\top A \leq c^\top$.*

Proof   As a consequence of Lemma 10 at most one of the alternatives can hold. First we consider the case that the primal program is feasible. If it is unbounded, the second alternative holds, thus we assume it is bounded. Let z_0 be a maximal lower bound. As z_0 is a lower bound there does not exist a pair $x, y \geq 0$ such that $Ax - y - b = 0$ and $c^\top x < z_0$. We examine the null space or kernel of the matrix

$$B := \begin{pmatrix} 0 & A & -I_m & -b \\ -1 & -c^\top & 0^\top & z_0 \end{pmatrix}.$$

If there exists some $(w_0, \tilde{x}^\top, y^\top, w_{n+1})$ in the null space of B such that $w_0 > 0, \tilde{x} \geq 0, y \geq 0, w_{n+1} \geq 0$, then, since z_0 is a bound, we necessarily must have $w_{n+1} = 0$, for otherwise $x := \frac{1}{w_{n+1}} \tilde{x}$ were feasible for the primal program and $c^\top x < z_0$. This however implies $A\tilde{x} \geq 0$ and $c^\top \tilde{x} = -w_0 < 0$, i.e. the primal program is unbounded, contradicting our assumption. Thus by Farkas' Lemma there exists $(\tilde{u}^\top, u_{m+1})$ such that $u_{m+1} < 0, \tilde{u}^\top A - c^\top u_{m+1} \geq 0, \tilde{u} \leq 0, u_{m+1} z_0 \geq \tilde{u}^\top b$. Now $u := \frac{\tilde{u}}{u_{m+1}} \geq 0$ and $u^\top A \leq c^\top$ as well as $u^\top b \geq z_0$. On the other hand z_0 is a maximal lower bound and $u^\top b$ is a lower bound of the primal program due to Lemma 10. Therefore we conclude $u^\top b \leq z_0$ and thus we have proven $u^\top b = z_0$.

Finally, we consider the case that the primal program is infeasible. In this case the matrix $(-b, A, -I)$ cannot have an element in its kernel that is non-negative and strictly positive in the first coordinate. By Farkas' Lemma there exists a $\tilde{u} \leq 0$ such that $\tilde{u}^\top b < 0$ and $\tilde{u}^\top A \geq 0$. Setting $u = -\tilde{u}$ we find that the dual program is unbounded if feasible.   □

That the last alternative may hold is illustrated with the following example:

Example 3

$$\begin{aligned} &\min -y_1 - y_2 \\ (P) \text{ subject to } \quad & y_1 - y_2 \geq 1 \\ & -y_1 + y_2 \geq 1 \\ & y_1 \ , y_2 \geq 0 \end{aligned}$$

$$\begin{aligned} &\max \quad x_1 + x_2 \\ (D) \text{ subject to } \quad & x_1 - x_2 \leq -1 \\ & -x_1 + x_2 \leq -1 \\ & x_1 \ , x_2 \geq \quad 0 \end{aligned}$$

Admittedly, our first alternative looks unnecessarily complicated and most textbooks on linear programming state it as

If the primal program is bounded, then the dual program is bounded as well and there are optimal solutions x^*, u^* such that $c^\top x^* = (u^*)^\top b$.

Our version is a bit different since it is not clear a priori, that the minimum of a primal program is actually attained and is not in fact only an infimum. However, it is an easy exercise now to fill this gap.

Exercise 26 Prove that the program dual to the dual program is the primal program. Hint: convert the dual program into standard form and dualize. Conclude that, if the primal program is feasible and bounded, then there exist optimal solutions x^* respectively u^* for both programs and furthermore: $c^\top x^* = u^{*\top} b$.

Exercise 27 Prove the following two statements.

(i) The dual of the program

$$\begin{aligned} &\min c^\top x \\ (P1) \quad & \text{subject to } Ax = b \\ & x \geq 0 \end{aligned} \qquad (4.7)$$

is the program

$$\begin{aligned} &\max u^\top b \\ (D1) \quad & \text{subject to } u^\top A \leq c^\top . \end{aligned}$$

(ii) The dual of the program

$$\begin{array}{ll} & \min c^{\top} x \\ (P2) & \text{subject to } Ax \geq b \end{array} \qquad (4.8)$$

is the program

$$\begin{array}{ll} & \max u^{\top} b \\ (D2) & \text{subject to } u^{\top} A = c^{\top} \\ & u \geq 0. \end{array}$$

In general the dual variables to inequality constraints have a restriction on their sign and dual variables to equality constraints have not. Along the same line the dual constraint to a variable with a constraint on its sign is an inequality whereas the dual constraint to an unconstrained variable is an equality.

Recalling that the purpose of this chapter is to consider Kruskal's algorithm from a primal-dual point of view, the following concept is central: variables of the primal program are classified according to whether the corresponding inequality of the dual is strict or an equality. In the former case we speak of a *positive slack*. In the latter case, i.e. when an inequality lacks slack, we call it *tight*.

Given a dual pair of linear programs satisfying the first alternative of Theorem 4, we consider a dual pair of optimal solutions x^{*}, u^{*}. We find

$$\begin{aligned} 0 = c^{\top} x^{*} - u^{*\top} b &= (u^{*\top} A + c^{\top} - u^{*\top} A) x^{*} - u^{*\top} b \\ &= u^{*\top} (Ax^{*} - b) + (c^{\top} - u^{*\top} A) x^{*}. \end{aligned}$$

As $x^{*}, u^{*} \geq 0$, $c^{\top} - u^{*\top} A \geq 0$ and also $Ax^{*} - b \geq 0$ we can conclude

Theorem 5 (Complementary slackness) *Let $x^{*} \in \mathbb{R}^{n}_{+}$ such that $Ax^{*} \geq b$ and $u^{*} \in \mathbb{R}^{m}_{+}$ satisfying $u^{*\top} A \leq c^{\top}$. Then x^{*}, u^{*} are a pair of optimal solutions of the primal respectively the dual program if and only if*

(i) $x^{*}_{i} \neq 0 \Rightarrow c_{i} = (u^{*\top} A)_{i}$,
(ii) $c_{i} > (u^{*\top} A)_{i} \Rightarrow x^{*}_{i} = 0$,
(iii) $u^{*}_{j} \neq 0 \Rightarrow b_{j} = (Ax^{*})_{j}$,
(iv) $b_{j} < (Ax^{*})_{j} \Rightarrow u^{*}_{j} = 0$.

Proof Necessity of the conditions has been proven in our consideration in the last paragraph. Conversely, the conditions imply

$$c^\top x^* = \sum_{i=1}^{n} c_i x_i^* = \sum_{i=1}^{n} (u^{*\top} A)_i x_i^*$$

$$= u^{*\top} A x^* = \sum_{j=1}^{m} u_j^*(Ax^*)_j = \sum_{j=1}^{m} u_j^* b_j = u^{*\top} b$$

and the assertion follows from Lemma 10.   □

4.4 The Greedy as Primal-Dual Algorithm

In the beginning of this chapter we presented the following linear program to solve the minimum spanning tree problem:

$$\min w^\top x$$

$$\text{subject to} \sum_{e \in \partial P} x_e \geq |P| - 1 \; \forall P \in \mathcal{P}$$

$$\sum_{e \in E} x_e = |V| - 1 \tag{4.9}$$

$$x \geq 0.$$

If we rewrite the last equation as two inequalities

$$\sum_{e \in E} x_e \geq |V| - 1 \text{ and } -\sum_{e \in E} x_e \geq -|V| + 1$$

and consider the $((|\mathcal{P}| + 1) \times |E|)$-Matrix A, with entries in $\{0, 1\}$, such that the rows are indexed with the partitions and the columns with the edges and that has a nonzero-entry if and only if $e \in \partial P$, then the above program has the form of our primal program and we can easily compute the dual program. We need a variable u_P for every non-trivial partition P and two u_+, u_- for the trivial one, yielding:

$$\max(|V| - 1)(u_+ - u_-) + \sum_{P \in \mathcal{P}}(|P| - 1)u_P$$

$$\text{subject to } (u_+ - u_-) + \sum_{e \in \partial P} u_P \leq w_e \qquad \forall e \in E$$

$$u_+, u_-, u_P \geq 0.$$

What do we gain from this? Contrary to the primal program, where a feasible solution is not completely obvious, here it is trivial to find one, e.g. by setting $u_P = 0$ for all non-trivial partitions and $(u_+ - u_-) = \min_{e \in E}\{w_e\}$.

A dually feasible solution u, by the complementary slackness Theorem 5, is optimal if and only if the *subgraph of tight edges* $\widetilde{G} = (V, \widetilde{E})$ where $\widetilde{E} := \{e \in E \mid (u_+ - u_-) + \sum_{e \in \partial P} u_P = w_e\}$ contains a spanning tree. Otherwise $\widetilde{G}$ is disconnected and its components form a non-trivial partition P_0 of the vertex set, such that

$$(u_+ - u_-) + \sum_{e \in \partial P} u_P < w_e \text{ for all } e \in \partial P_0.$$

Thus, if we update u_{P_0} from zero to $u_{P_0} := \min_{e \in \partial P_0} w_e - (u_+ - u_-) - \sum_{e \in \partial P} u_P > 0$ then u remains feasible. This yields a new dually feasible solution and we may proceed. In each iteration the number of components of the graph of tight edges decreases, and thus the algorithm terminates in at most $|V| - 2$ iterations.

The main steps of the proposed algorithm thus are

ALGORITHM PrimalDualKruskal

```
while nrOfComponents() > 1:
    UpdateDuals()
    ComputePartition()
```

Software Exercise 28 There is a nice geometric interpretation which we learned from M. Jünger and W. Pulleyblank [26] for the case of a complete graph on a set of points in the plane, where edge weights are the Euclidean distances between points.

Go to the directory 04-LPDuality and load the algorithm PrimalDual Kruskal.alg and a graph, say, PD_Kruskal4.cat. As soon as you start the algorithm all edges will disappear. We start with the trivial partition where each single point is a class of its own.

To find the minimum distance between the classes of the partition we grow moats around each point until two of the moats meet. The radius of the moats is the new value of our dual variable and the edges perpendicular to each pair of meeting moats enter our graph of tight edges in the second window. In the next step we compute the components of that graph to determine the new partition and iterate until there is only one component left.

Theorem 6 *When the algorithm Primal Dual Kruskal terminates, then $T := supp(x)$ is a spanning tree of G minimizing $\sum_{e \in T} w(e)$.*

Proof Throughout the algorithm we have a complementary pair (x, u) where x is the incidence vector of a forest and u is dually feasible. In each iteration of the while loop the size of the partition decreases by at least one. Thus the while loop is executed at most $n - 1$ times. By construction, T is a tree and $x := \chi(T)$ is feasible for the primal program. Similarly u is dually feasible and u and x fulfill the assumptions of Theorem 5 and x is an optimal solution of the linear program for Problem 2. $\qquad\square$

4.5 The Base Polytope

This section is an extended exercise meant to study the proof of Theorem 3. Again this proof is of an intrinsic matroid theoretic nature. Once understood you will appreciate that it becomes simpler if you do it on a higher level of abstraction.

In order to generalize Theorem 3 to matroids, first we have to find the analogs to spanning trees and to partitions. So, let $M = (E, C)$ be a matroid.

Definition 12 An independent set B such that $B \cup \{e\}$ is not independent for all $e \in E \setminus B$ is called a *basis* of M.

Recall from Exercise 20 that the bases of a graphic matroid are the spanning trees. Clearly, the naming in a vector matroid coincides with linear algebra. From that we can also generalize the notion of a dimension. The reason is the following.

Theorem 7 *Let B_1, B_2 be two bases in a matroid. Then $|B_1| = |B_2|$.*

Proof Let B_1 be a basis of smallest cardinality in M. Assume there exists a basis of larger cardinality. Then we may choose a basis B_2 such that $|B_2| > |B_1|$ and $|B_2 \cap B_1|$ is maximal. Let $g \in B_1 \setminus B_2$ and C_g the elementary circuit of $B_2 \cup g$. Choose some $f \in C_g \setminus B_1$. Then $B_3 := (B_2 \setminus f) \cup g$ is independent and $|B_3| = |B_2| > |B_1|$ but $|B_3 \cap B_1| = |B_2 \cap B_1| + 1$. Thus any basis containing B_3 contradicts the maximality of B. □

To simplify notation we introduce the following.

Definition 13 For $S \subseteq E$ let $M[S] = M \setminus (E \setminus S)$ denote the *matroid induced on S*.

Recall that in the proof of Theorem 3 we had to glue together spanning trees of the graph induced on the non-trivial class V_1 of a partition and spanning trees of G/V_1. Any such pair yielded a spanning tree of G. Translated into matroid language we have:

Theorem 8 *Let $S \subseteq E$, B_1 be a basis of $M[S]$ and B_2 be a basis of M/S. Then $B_1 \cup B_2$ is a basis of M.*

Proof B_2 is a basis of M/S and thus, no circuit in $B_2 \cup S$ may intersect B_2. Since B_1 itself does not contain a circuit, $B_1 \cup B_2$ must be independent.

Assume there exists some $f \in E \setminus (B_1 \cup B_2)$ such that $B_1 \cup B_2 \cup f$ is independent. Then, $f \in S$ contradicts B_1 being a basis of $M[S]$. Thus, $f \notin S$. Let $\widetilde{C}_f$ be the elementary circuit $B_2 \cup f$ with respect to M/S. By definition of a contraction minor we can complete this to a circuit C_f of M using some elements from S. We choose these elements such that $|S \setminus B_1|$ is as small as possible. Let $g \in (C_f \setminus B_1) \cap S \neq \emptyset$ and C_g be the elementary circuit $B_1 \cup g$. This is a circuit in M. Applying circuit elimination we find $C_2 \subseteq C_f \cup C_g \setminus g$. Necessarily we must have $C_2 \setminus S = \widetilde{C}_f$ contradicting the choice of C_f. □

The generalization of the dimension is the following:

Definition 14 For $S \subseteq E$ we define the *rank of S* to be

$$r_M(S) := \max\{|I| : I \subseteq S \text{ is independent}\}. \tag{4.10}$$

With this terminology we get as an immediate corollary from Theorem 8:

Corollary 2 *For $T \subseteq E \setminus S$ we have $r_{M/S}(T) = r_M(T \cup S) - r_M(S)$.*

□

The next object to study are the *closed subspaces* of a matroid, those sets which are maximal.

Definition 15 A set $H \subset E$ is *closed in M* or a *flat*, if

$$r_M(H \cup e) > r_M(H) \text{ for all } e \in E \setminus H. \tag{4.11}$$

What are the flats of a graphic matroid? They are those sets $H \subset E$, such that for any edge from $E \setminus H$, $H \cup \{e\}$ has fewer components than H. Thus, if $P = V_1 \dot\cup \ldots \dot\cup V_k$ is the partition into components induced by H, then $H = E \setminus \partial P$. If we recall the inequalities from $ST(G)$

$$\sum_{e \in \partial P} x_e \geq |P| - 1 \text{ and } \sum_{e \in E} x_e = |V| - 1, \tag{4.12}$$

and take into account that $r(E \setminus \partial P) = |V| - |P|$, we can generalize these to

$$\sum_{e \in H} x_e \leq r_M(H) \text{ and } \sum_{e \in E} x_e = r_M(E). \tag{4.13}$$

With these preparations the proof of Theorem 3 easily is transferred to the setting of matroids. We leave the details as an exercise to reader.

Theorem 9 *Let $M = (E, \mathcal{C})$ be a matroid. For $S \subseteq E$ we use the abbreviation $x(S) := \sum_{e \in S} x_e$. Let*

$$ST(M) := \{x \in \mathbb{R}_+^E : x(S) \leq r_M(S) \text{ for all } S \subseteq E \text{ and } x(E) = r_M(E)\}$$

and

$$P(M) := \text{conv}(\{\chi_B : B \subseteq E \text{ is basis of } M\}).$$

Then

$$ST(M) = P(M).$$

We call this polyhedron the base polytope *of M.*

Proof   Left as an exercise.   □

Exercise 29 Prove Theorem 9.

Exits

Linear programming is a topic for at least one full course and we could address only the very basics here. In particular we did not discuss algorithmic issues of general linear programs. There are several classical textbooks [14, 9] that in particular discuss the simplex method which probably is still the most frequently applied algorithm for solving linear programs. A drawback of this method is that no polynomial running time bound is known and for basically all implementations (pivoting rules) classes of examples are known that do not admit such a polynomial bound. There are polynomial time algorithms for linear programming though, but they have their origin in continuous methods for non-linear programming [42].

For our purposes linear programming will be more a theoretical tool to develop combinatorial algorithms in particular of the primal-dual-type. Alexander Schrijver wrote a comprehensive treatment of this approach [39] which is heartily recommended to anybody seriously interested in combinatorial optimization.

We will not consider general linear programming problems but will conclude this chapter with a typical representative, a blending problem, which we will give as an exercise.

Exercises

Exercise 30 A portfolio manager shall make an investment of 70 Mio. $. The risk of the different types of investments is estimated by a numerical value ranging from 0 (no risk) to 100 (very high risk). The following products are available:

Type	Growth Stock	Blue Chips	Foreign Stock	Bonds	Cash
Rate of Return	14.25%	8.75%	11.25%	7.75%	4.45%
Risk	90	50	70	30	0

The managers objective is to maximize the rate of return subject to the following constraints. At least 3.5 Mio. $ in cash must be held back for flexibility, at most 80% may be invested in stocks (growth stock, blue chips, foreign stock). At least 10% of the stock shall be foreign stock and risk shall on average not exceed a value of 50.

Chapter 5
Shortest Paths

5.1 Introduction

After the abstract algebraic material of the last chapter, this one will be easier and more concrete. We shall deal with the problem of determining a shortest path in a weighted graph. This everyday problem does not only occur in its most obvious form, as we already realized in Chap. 1.

Classically, in combinatorial optimization one distinguishes between the following variants of the problem.

Problem 4 Let $D = (V, A)$ be a directed graph and *length*: $A \to \mathbb{Z}$ a weight function on the arcs. Consider the shortest path problem variants:

(i) st-path: Let $s, t \in V$. Find a shortest path from s to t.
(ii) s-paths: Let $s \in V$. Find shortest paths from s to all $v \in V$.
(iii) distances: Determine the distance; i.e., the length of a shortest path between any pair of vertices $u, v \in V$.

Remark 6 In general we will ignore the question whether there exists a path at all. We always can enforce the existence of such a path by adding all possible arcs not already present in A and assigning them a length of infinity.

One might consider the problem of an st-path most natural, but we will deal mainly with s-paths as all known methods for solving the first variant also solve the second, at least partially. They simply compute all the s-paths, stopping when the vertex t is encountered.

Why do we insist on digraphs in the definition of the problem and do not simply replace each edge in an undirected graph by two arcs in opposite directions but of the same weight? Actually, this works fine as long as all weights are non-negative, but in certain applications they are not. If we admit negative arc lengths, the minimum length of a chain between any pair of nodes becomes unbounded. The reason is that a pair of arcs of negative length creates a circuit of negative length. The presence of negative circuits apparently makes the problem intractable, it becomes $\mathcal{NP}$-complete (meaning that it belongs to a class of problems where one does not know an efficient algorithm for any of its members, but where an efficient algorithm

W. Hochstättler, A. Schliep, *CATBox*, DOI 10.1007/978-3-642-03822-8_5,
© Springer-Verlag Berlin Heidelberg 2010

for one member can be converted to an algorithm solving all problems in the class efficiently). Arcs with negative length alone do not cause similar problems and we will learn a method to solve problems with negative arc lengths, later on. Summarizing, we distinguish between problems with

  (i) non-negative weights,
 (ii) without negative circuits and with
(iii) arbitrary weights.

We will be mainly concerned with the first variant and will not discuss the last.

5.2 Dijkstra's Algorithm

The most popular method to solve the shortest path problem is due to E.W. Dijkstra. The method works properly only if all arc lengths are non-negative. Because of the practical relevance of this problem there exists a huge variety of implementations improving the complexity under certain circumstances. We will not discuss this in any detail here, but rather refer to textbooks, for example [1].

The exact problem the algorithm solves is the following.

Problem 5 Let $D = (V, A)$ be a directed graph, $w : A \to \mathbb{Z}_+$ a non-negative weight function on the arcs and $s \in V$. Find a shortest path from s to v for all $v \in V$.

To get an idea of Dijkstra's method imagine the following mind experiment. We all know that the light travels on shortest paths. How does light behave, if it cannot follow the bee-line, say a signal is broadcasted in a net of glass fibers. In the beginning there is a distinguished node s that activates the experiment by sending the signal to all its neighbors. A node that has not received a signal yet is in waiting mode. As soon as a node that is in waiting mode receives the first signal, it assigns a marker to the incoming wire transmitting the signal, and sends a signal to all its other neighbors. A node that has broadcasted a signal disconnects. After this procedure has finished we can determine a shortest path from s to a node v by backtracking the markers of the vertices, starting at v until we reach s, thus traversing the shortest path backwards.

To turn this intuitive approach into an algorithm we keep a distance label `dist[v]` for all $v \in V$ and a pointer `pred[v]`. When the algorithm has terminated `dist[v]` will contain the length of a shortest path from s to v. To determine the shortest path from s to v we have to backtrack along the pointers `pred[v]`, which correspond to the markers on the incoming wires in the previous mind experiment.

During the whole procedure, any vertex is in exactly one of the following three states: It is *unvisited, visited or processed*. We will organize this in a set W containing the visited vertices. A vertex is unvisited, if it was not yet added to W and processed if it is a former member of W. In our mind experiment the processed vertices are those nodes that have disconnected, the distance label corresponds to its point of time of broadcast, a visited node is a node that has not received a signal yet but where at least one of its neighbors already has disconnected. Their distance label is the earliest point of time where the signal from one of its disconnected neighbors will arrive, `pred[v]` points to the origin of the received signal. The unlabeled vertices have a distance label of ∞.

ALGORITHM Dijkstra

```
  s = PickVertex()
  dist[s] = 0
  W.Add(s)

5 while W.IsNotEmpty():
      v = pickMinimal(W,dist)
      W.Remove(v)
      for w in Neighborhood(v):
          if dist[w] == gInfinity:
10            W.Add(w)
          if dist[w] > dist[v] + length[(v,w)]:
              dist[w] = dist[v] + length[(v,w)]
              pred[w] = v
```

Theorem 10 *Let $G = (V, A)$ be a connected digraph, $s \in V$. Dijkstra's Algorithm computes the lengths of shortest s-v-paths for all $v \in V$. Backtracking the* pred-*pointers yields shortest paths.*

Proof First note that any vertex is added to W exactly once as the graph is connected. In every iteration of the while-loop one vertex is removed from W and is processed. Thus the algorithm is finite. Let S denote the set of processed vertices, i.e. those that have been removed from W. We show by induction on $|S|$ that at any time of the algorithm for all vertices in S holds: backtracking the pointers `pred[v]` yields a shortest s-v-path since whenever `pred` is changed in the algorithm it points from a vertex in W to a vertex in S.

$|S| = 1$: Here, $S = \{s\}$ and there is nothing to show since the graph does not contain negative circuits.

$|S| > 1$: Let v be the last vertex that has been added to S. Let $s = v_0$, $v = v_k$ and $v_0 v_1 \ldots v_k$ where $v_{i-1} = \text{pred}[v_i]$ be the sv-path given by the `pred`-pointers. Denote the arc set of this path by P. From the statement in the algorithm where `dist[`v_k`]` is set, we inductively conclude that $\text{length}(P) := \sum_{i=1}^{k} \text{length}$ $[(v_{i-1}, v_i)] = \text{dist}[v]$. Assume, that $\widetilde{P} = e_1, \ldots, e_{k'}$ were the arc sequence of an s-v-path $v_0' v_1' \ldots v_{k'}'$ where $s = v_0'$ and $\text{length}(\widetilde{P}) < \text{dist}[v]$ (see Fig. 5.1).

Let i_0 denote the largest index such that $v_{i_0}' \in S \setminus \{v\}$ and set $e_{i_0} := (v_{i_0}', v_{i_0+1}')$. While v_{i_0}' is processed in the algorithm, it is checked whether $\text{dist}[v_{i_0}'] + \text{length}[e_{i_0}] < \text{dist}[v_{i_0+1}']$. Thus after this step we must have

$$\text{dist}[v_{i_0+1}'] \leq \text{dist}[v_{i_0}'] + \text{length}[e_{i_0}]. \tag{5.1}$$

By inductive assumption $\text{dist}[v_{i_0}']$ is the length of a shortest s-v_{i_0}'-path. As, furthermore, the length function is non-negative, we conclude $\text{dist}[v_{i_0}'] + \text{length}[e_{i_0}] \leq \text{length}(\widetilde{P}) < \text{length}(P) = \text{dist}[v]$. Therefore, before choosing v from W, according to the rule of the algorithm, we should have chosen $v_{i_0+1}' \notin S$, a contradiction. $\qquad\square$

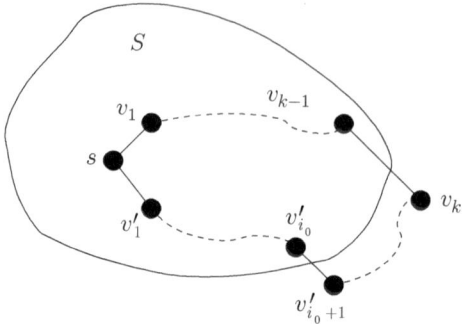

Fig. 5.1 In the proof of Dijkstra's algorithm

5.3 Some Considerations about Implementations

In the inner **for**-loop we evaluate each arc exactly once. For each arc the evalua-
tion requires constant time. Since we have to handle each arc at least once, there
is hardly a chance for an improvement and the running time of $O(|A|)$ is opti-
mal for that loop. A lot more interesting is the implementation of `pickMinimal`
(`W,dist`). If in each iteration we do this search in a naive way, as it is implemented
in the prologue of `Dijkstra.alg`, we get a running time of $O(|V|^2)$. This can
be overcome by keeping the vertices in a sorted list with a suitable data structure
(e.g. a heap) where we account in total $O(|V|\log(|V|))$ for initial construction and
updates. In our implementation of the different variants of Dijkstra's algorithm we
use a *priority queue*, which keeps the elements of the list in a non-increasing order
and allows to update values. We can add an element to that list or make an update
in $O(\log|V|)$. This results overall in the above complexity. We implemented W as a
priority queue in `DijkstraPQ.alg`. As you see, the main difference is that, here
we also have to update the position of a vertex w in the queue whenever `dist[w]`
is changed.

If `length` is bounded by a constant C and all weights are integers, it is possible
to "sort" in linear time using a "bucket sort" (each number gets its own bucket). This
trick can be adapted to the general situation using "logarithmic buckets" resulting in
a complexity of $O(|V|\log(M))$ where $M = \max\{$`length[e]`$\,|e \in A\}$ (see e.g. [1]
for details).

5.4 Special Label Setting Methods for the Computation
of Shortest Paths Between Two Vertices

Returning to our definition of s-t-paths in Problem 4 we can solve this problem by
letting Dijkstra's algorithm run until vertex t has been processed. Another idea that
usually reduces the number of vertices to be processed is to let Dijkstra's method run
simultaneously from both terminal nodes s and t. Clearly, we have to work on arcs
in reverse direction when searching for shortest paths to t. The method terminates

when the sets of vertices labeled permanently from s and from t have a vertex in common.

Further constraints on the class of input graphs can lead to faster computations. When we compute shortest paths to choose a travel route in real life, we usually do not compute shortest paths in all directions until we hit our destination. For example, it seems to be quite unlikely that it is helpful to consider routes via Brazil, when we want to travel from Mexico to Canada. Indeed in this case we can modify our method by additionally considering distances as the crow flies.

In general, we say that the function w is *geometric* if there exists a lower bound distance: $V \to \mathbb{R}$ for the length of all shortest paths to the destination t, i.e., such that distance[v] $\le$ dist[v,t] for all vertices $v \in V$, such that this bound satisfies the triangle inequality, i.e.

$$\forall v, w \in V : \text{distance[w]} + dist(w, v) \ge \text{distance[v]}.$$

In the case of points in the plane we can derive such a function from *coordinates* (v_x, v_y) of the vertices v such that for all arcs $e = (u, v)$:

$$\text{length[(u,v)]} \ge \sqrt{(v_x - u_x)^2 + (v_y - u_y)^2},$$

distance is chosen as the linear distance

$$\text{distance[v]} := \sqrt{(v_x - t_x)^2 + (v_y - t_y)^2}.$$

In our geometric variant of Dijkstra's algorithm we pick a vertex that is minimal not just with respect to its label—the distance from s in the graph—but to the sum of its label and the distance to the terminal vertex t as given by the Euclidean distance in the plane. This favors vertices that lie in the direction of the destination (Fig.5.2).

ALGORITHM FindPathEuclid

```
while v != t:
    for w in Neighborhood(v):
        if dist[w] == gInfinity:
            dist[w] = dist[v] + length[(v,w)]
            pred[w] = v
            PQ.Insert(w,dist[w]+distance[w])
        elif dist[w] > dist[v] + length[(v,w)]:
            dist[w] = dist[v] + length[(v,w)]
            pred[w] = v
            PQ.DecreaseKey(w,dist[w]+distance[w])

    v = PQ.DeleteMin()

ShowPath(s,t)
```

Theorem 11 *Algorithm Euclidean Shortest Path computes a shortest path from s to t.*

Proof The proof is an exact copy of the proof of Dijkstra's original algorithm except that (5.1) is replaced by

$$\texttt{dist}[v'_{i_0+1}] + \texttt{distance}[v'_{i_0+1}] \le \texttt{dist}[v'_{i_0+1}] + \mathit{dist}(v'_{i_0+1}, v_k)$$
$$+\texttt{distance}[v_k] < \texttt{dist}[v_k] + \texttt{distance}[v_k]$$

□

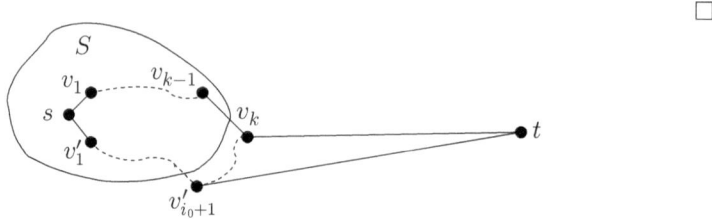

Fig. 5.2 In the proof of the Euclidean modification of Dijkstra's algorithm

Software Exercise 31 It is obvious that the above implementation can be superior to Dijkstra's original algorithm only if we stop our search as soon as the destination vertex is reached. Implementing this approach based on a priority queue is FindPathEuclid.alg. Using a priority queue we have to insert that element into our list according to its distance. Also we have to update its position in the list, whenever we change its distance.

We suggest that you let the algorithm Dijkstra.alg operate on 11x11grid.cat, where the objective is to find a shortest path from the center to the corner in the lower right. It is easy to see how the queue spreads in concentric circuits like a wave. When vertex t is encountered, all vertices have been visited and all but one have been explored. Note, that due to the numbering of the vertices all paths in our shortest path tree have at most one turn.

Now compare this to the result of FindPathEuclid.alg. After three vertices have been processed, we ignore the other two vertices at dist 50 from s and process vertex 73 instead, which has a dist of 100 to s because its Euclidean distance to t is more than 100 units smaller than that of vertices 50 and 62. The queue now spreads more in an elliptic fashion and the resulting shortest path zig-zags. Furthermore, only about 30% of the vertices are explored.

Finally you can compare this to the approach used in TwoSources.alg where concentric circuits are grown around the source and the destination simultaneously and the algorithm terminates, when a vertex becomes permanently labeled from both sides. It becomes apparent that we can save some computation here as well, at least in the geometric case.

5.5 Label Setting Versus Label Correcting

Dijkstra's algorithm is called a label setting algorithm since the label of a vertex that is processed will never be changed later in the algorithm.

The label setting is only feasible since we, due to non-negativity of the arc lengths, know in advance that prolonging a path cannot make it shorter. But even if there are negative arc lengths we are not completely lost. If there is no circuit of negative length we can use a label correcting method to solve it. We consider here the method of Bellman and Ford which is fairly easy to implement. Since any sorting is avoided, the method is, although inferior in theory, competitive to Dijkstra's method for standard instances.

ALGORITHM BellmanFord

```
 1  s = PickVertex()
    dist[s] = 0
    pred[s] = None
    Q.Append(s)

 6  while Q.IsNotEmpty():
        v = Q.Top()
        for w in Neighborhood(v):
            if dist[w] > dist[v] + length[(v,w)]:
                pred[w] = v
11              dist[w] = dist[v] + length[(v,w)]
                if not Q.Contains(w):
                    Q.Append(w)
```

Theorem 12 *If $D = (V, A)$ is a digraph and* length $: A \to \mathbb{Z}$ *a weight function on the arcs, such that there is no directed circuit C of negative total length $\sum_{a \in C}$ length$[a] < 0$ in D, then the above algorithm computes the shortest paths from s to all other vertices.*

Proof First we show that the algorithm terminates. The reason is that whenever the distance of a vertex dist$[v]$ is updated, backtracking the pred-pointers yields a directed s-v-path of length dist$[v]$, since there are no negative circuits. There are only finitely many paths in the graph and for every vertex the function dist$[v]$ is monotonically decreasing throughout runtime, hence the algorithm is finite.

When the algorithm terminates we have dist$[u]+$length$[a] \geq$ dist$[v]$ for all $a = (u, v) \in A$. Now let $s = v_0, a_1, v_1 \ldots, a_k, v_k = v$ be the vertex arc sequence of some s-v-path $\widetilde{P}$. Then length$(\widetilde{P}) = \sum_{i=1}^{k}$ length$[a_i] \geq \sum_{i=1}^{k}$ dist$[v_i] -$dist$[v_{i-1}] =$ dist$[v] -$dist$[s] =$ dist$[v]$. Thus, the path defined by the pred pointers is shortest. $\square$

Remark 7 The running time of a general version of a label correcting algorithm, i.e. pick some arc, check for dist$[u] +$ length$[a] \geq$ dist$[v]$ and update if necessary, is not bounded by a polynomial. The Bellman-Ford algorithm that uses a First-in-First-out Queue has a running time of $O(|V||A|)$ in the worst case. This is an immediate consequence of the following lemma.

Theorem 13 *If $u_1, \ldots, u_k$ are the vertices in the queue Q then the p_i number of arcs of the paths from s to the u_i derived from backtracking is monotonically increasing and in each iteration for all i, j it holds $|p_i - p_j| \leq 1$.*

Proof Clearly, the assertion is correct in the beginning. Whenever we add some vertex v to the queue while scanning the first vertex u_1 from the queue, the path to v will use one more arc than the path to u_1, implying the assertion inductively. $\square$

Software Exercise 32 If we compare the behavior of the Bellman-Ford algorithm, called `BellmanFord.alg` here, with Dijkstra on the graph `11x11grid.cat` we do not see any difference. The computations are identical, but we save the cost of sorting, which makes Bellman-Ford superior here.

The worst-case running time of the Bellman-Ford algorithm instead, even on a planar example with Euclidean distances, is demonstrated with the graph `Bellman FordWC.cat`. This graph consists of a path of length $O(n)$, "detours using few arcs" for each vertex on the path and a net of $O(n)$ at the end. Check the example to see that each of the vertices in the net is relabeled $O(n)$ times. The running time is $O(n^2) = O(|V||A|)$. The latter equation holds for planar graphs (see, e.g., [18]). If you run this example with Dijkstra's algorithm, the difference between label setting and label correcting should become clear.

On the other hand try both algorithms on the graphs `NegCircuit.cat` and `NegArc.cat`. The first example contains a negative circuit with the effect that Bellman-Ford does not terminate and Dijkstra produces garbage. The latter is also true for the second example. After termination we see the correct shortest path from 1 to 8 but its length is claimed to be about 824 (move the mouse on vertex 8 and check the info line). You can also see that something is wrong from the fact that there still is a temporarily labeled vertex 5 (it has been labeled for a second time). If you run Bellman-Ford on it you will find that the length of the path is 684. As last example we recommend `NegArc2.cat` where Dikstra's algorithm produces the wrong path.

The shortest path computations may be considered as the prototype of a *dynamic program*. Because of the importance of this concept we will discuss Application 4 in more detail.

Before doing so we will present an algorithm that solves the all-pairs shortest path problem. Due to its nature there is very little visualization in our implementation. Let A denote the $(|V| \times |V|)$-matrix, having the entries $A_{i,j} = $ `length[(i, j)]`, if (i, j) is an arc of G and ∞ otherwise. Trivially, A contains the lengths of all shortest paths that use a single arc between all pairs of vertices. Now, we consider the operation

$$C_{i,j} := \min_{k=1}^{|V|}\{A_{i,k} + A_{k,j}\}.$$

If we define the matrix C this way, then $C_{i,j}$ will contain the length of a shortest path from i to j that uses at most two arcs. We can iterate this and define the operation $B \bowtie A$ for two quadratic $(n \times n)$ matrices

$$(B \bowtie A)_{ij} := \min_{k=1}^{n}\{B_{i,k} + A_{k,j}\}.$$

This results in the following algorithm, which can be run from the python interpreter. A is an example weight matrix, 999 represents a non-edge.

ALGORITHM `Allpairs.py`

```
def Allpairs(A):
    n = len(A)
    B = A
    for l in range(n-2):
        for i in range(n):
            for j in range(n):
                B[i,j] = min(B[i,:] + A[:,j])
    return B
```

Theorem 14 *Let $D = (V, A)$ be a digraph and* `length` $: A \to \mathbb{Z}$ *a weight function, such that the digraph does not contain a directed circuit of negative length. The above method computes the length of a shortest u-v-path for all pairs of vertices $u, v \in V$.*

Proof We have seen above that after the i-th iteration the matrix B has the lengths of the shortest paths using at most $(i + 1)$ arcs as entries. No path in G can use more than $|V| - 1$ arcs and, thus, the claim follows.   □

The work load for the operation $B \bowtie A$ in a naive implementation is the same as it is for matrix multiplication, namely $O(|V|^3)$. This gives an overall running time of $O(|V|^4)$.

Remark 8 If the algorithm is run on a digraph with arbitrary arc weights, then this digraph contains a negative directed circuit if and only if after termination of the algorithm we have $B \bowtie A \neq B$.

Floyd and Warshall observed that the above method can be made significantly faster, namely $O(|V|^3)$, if instead of all paths of length at most k only those are considered that use only nodes from the set $\{1, \ldots, k - 1\}$ as inner nodes.

Proof This is left as an exercise.   □

Exercise 33 Consider the following algorithm:

ALGORITHM `Floyd-Warshall`

```
def FloydWarshall(A):
    n = len(A)
    B = A
    for l in range(n):
        for i in range(n):
            for j in range(n):
                B[i,j] = min(B[i,j], B[i,l] + B[l,j])
    return B
```

Prove that the Floyd-Warshall computes the distance between all pairs of vertices in $O(n^3)$.

5.6 Detecting Negative Circuits

An algorithm that only works properly on a restricted set of instances is of little practical help if one does not know how to tell good instances from bad ones, we call this the problem of *recognizing* good instances. In this section we will discuss a way to modify the algorithm of Bellman and Ford such that it detects a negative circuit if one exists.

Recall from Theorem 13 that we can associate a label with each vertex in the queue representing the number of arcs in the path encoded by the `pred`-pointers, so that these labels are monotonically increasing and differing by at most one throughout the algorithm.

Just as in the Floyd-Warshall algorithm a negative circuit will yield a walk from s to some vertex v using $|V|$ edges that is shorter than any s-v walk using at most $|V| - 1$ edges.

Lemma 11 *Let $G = (V, A)$ be a directed graph,* `length` $: A \to \mathbb{Z}$ *a weight function and $s \in V$. There exists a directed circuit of weight $\sum_{e \in C}$ `length[e]` < 0, that can be reached from s on a directed path, if and only if there exist $v \in V$ and an s-v walk P of length $|V|$ such that for all s-v walks $\widetilde{P}$ of length at most $|V| - 1$ holds:*

$$\sum_{a \in P} \text{length[a]} < \sum_{a \in \widetilde{P}} \text{length[a]}. \tag{5.2}$$

Proof If there exists such a directed walk P it cannot be acyclic by Exercise 3. Let C be a circuit in P. Then $\widetilde{P} := P \setminus C$ is an s-v walk using fewer arcs and thus

$$\sum_{a \in C} \text{length[a]} = \sum_{a \in P} \text{length[a]} - \sum_{a \in \widetilde{P}} \text{length[a]} < 0. \tag{5.3}$$

On the other hand let C be a circuit of minimum total weight less than 0. Assume that the claim is false for all vertices on C and let `dist[v]` denote the length of a shortest s-v-walk using at most $|V| - 1$ edges. Then by assumption for edges $e = (u, v) \in C$

$$\text{dist[u]} + \text{length[(u,v)]} \geq \text{dist[v]}. \tag{5.4}$$

Summing up these equations yields $\sum_{a \in C}$ `length[a]` ≥ 0, a contradiction. $\square$

Thus, in order to detect a negative circuit we proceed in steps and partition the queue into two. NEXT contains the vertices where the actual paths from s use at most `count` edges and NOW those using `count` $+ 1$ edges. If NEXT is completely processed we increment the `counter` and move all vertices from NEXT into NOW. If after the $|V| - 1$-st iteration NEXT is not empty, it must contain an element v such that there exist an s-v-walk using $|V|$ edges that is shorter than any such walk using less edges. Backtracking its pointers we have to run into a negative circuit. Therefore we leave marks to encounter a vertex `root` that is visited twice. Backtracking from `root` delivers the negative circuit.

Altogether we derive at the following algorithm

ALGORITHM NegativeCircuits

```
s = PickVertex()
dist[s] = 0
pred[s] = None
count = 0
```

```
5
  NEXT.Append(s)

  while count < G.Order() and NEXT.IsNotEmpty():
      if NOW.IsEmpty():
10        while NEXT.IsNotEmpty():
              v = NEXT.Top()
              NOW.Append(v)
              count += 1
      while NOW.IsNotEmpty():
15        v = NOW.Top()
          for w in Neighborhood(v):
              if dist[w] > dist[v] + length[(v,w)]:
                  pred[w] = v
                  dist[w] = dist[v] + length[(v,w)]
20                if not NEXT.Contains(w):
                      NEXT.Append(w)

  if count == G.Order():
      t = NEXT.Top()
25    marked[t] = 1
      while not marked.QDefined(pred[t]):
          t = pred[t]
          marked[t] = 1
      root = t
30    while pred[t] != root:
          circuit.AddEdge((pred[t],t))
          t = pred[t]
      circuit.AddEdge((pred[t],t))
```

Software Exercise 34 You can run `NegativCircuit.alg` on `NegCircuit.cat` or `NegCircuit2.cat`. The blue vertices represent `NOW` and the red `NEXT`. If the graph does not have a circuit of negative total weight the algorithm computes a shortest path tree. Otherwise, the yellow vertices in the end are those which become marked during the backtracking and the yellow edges build a circuit of negative weight.

If you have time to run an algorithm for half an hour you can search for a negative edge in `11x11neg.cat`. We interpret an undirected graph as a directed graph with edges in both directions. Thus, a negative edge, yields a negative circuit. As the graph has 121 vertices, the **while**-loop has to be executed that often.

5.7 Application

We now turn to Application 4 which we briefly introduced in the first chapter. Formally we can describe the problem as follows:

Problem 6 Let $B = b_1, \ldots, b_k$ and $D = d_1, \ldots, d_l$ be two words on the alphabet $\{A, C, G, T\}$. Words can be transformed into one another with the the following operations:

- *Insertion:* Inserting a letter at any position is done for a cost of $\alpha \geq 0$.
- *Deletion:* Deletion of an arbitrary letter costs $\beta \geq 0$.
- *Mutation:* Finally, it is possible to change a letter b_i into d_j at a cost of $g(b_i, d_j)$, where we assume that the letter does not change its position during this operation.

We assume that the costs of the mutations satisfy the triangle inequality, i.e. $g(x, y) + g(y, z) \geq g(x, z)$. Find the (some) cheapest way to transform B into D!

If we consider the infinite graph of all possible words over this alphabet, we can consider this problem as a shortest path problem on an infinite graph. Two vertices are linked by an arc if they can be transformed into one another by a single insertion, deletion or mutation. Even though the problem is defined on an infinite graph we can solve it by exploiting that a shortest path between any pair of words will use at most $k + l$ arcs (delete the first word and insert the second one letter for letter). Obviously, there is no point in deleting or mutating an inserted letter or deleting or mutating a mutated letter (triangle inequality). Thus we can restrict our attention to words of a length of at most $\max\{k, l\}$ and run Dijkstra's algorithm.

Making even better use of the problem structure, we will present a dynamic program. Its key idea resembles the operation in the matrix algorithm. This program exploits the fact that the exact order of the operations that transform one word into the other does not matter. Starting from a path that deletes all letters and inserts the second word we search for shortcuts starting at the point where B has just been completely deleted.

For $1 \leq i \leq k, 1 \leq j \leq l$ we denote by $f(i, j)$ the cost of a cheapest transformation of the (partial) word $b_1, \ldots, b_i$ into the (partial) word $d_1, \ldots, d_j$. Then

- $f(0, j) = \alpha j$,
- $f(i, 0) = \beta i$, and
- $f(i, j) = \min\{f(i - 1, j - 1) + g(b_i, d_j), f(i, j - 1) + \alpha, f(i - 1, j) + \beta\}$.

The last line results from distinguishing between b_i mutating to d_j, d_j being inserted or b_i deleted. One of these has to be the last operation in a cheapest transformation.

Using the recursion above in $O(kl)$ we can compute $f(k, l)$ by table filling. Such a computation is called *dynamic programming*.

5.8 An Analog Computer and LP Duality

Consider an instance of the shortest path problem with nonnegative arc weights as a net in the classical sense, consisting of inelastic ropes and knots. The length of the rope between two knots reflects the length of the arc between these two vertices. A shortest path between two vertices s and t is now easily found. Take these two vertices and pull them apart as far as possible. Any path from s to t that uses only taut rope is a shortest path.

Let us describe this analog computer using our algebraic formalism introduced in Chap. 4.

Problem 7 Let $D = (V, A)$ be a digraph and $w : A \to \mathbb{Z}$ a weight function and $s, t \in V$. For each vertex $v \in V$ we introduce a variable d_v and consider the task to maximize $d_t - d_s$ subject to

$$d_v - d_u \leq \texttt{length}[e]$$

for all arcs $e = (u, v)$.

If we denote by $\chi(v)$ the characteristic function of a vertex v, by d the vector $(d_s, \ldots, d_t) \in \mathbb{Z}^{|V|}$, by $(\texttt{length}[e])_{e \in A} \in \mathbb{Z}^A$ the weight vector and by B the vertex arc incidence matrix, we can write this problem as

$$\max d^\top(\chi(t) - \chi(s))$$
$$\text{subject to} \qquad d^\top B \qquad \leq w$$

Note, that the set of optimal solutions is invariant under translation, i.e. adding the same constant to each component does change neither feasibility nor optimality. Thus, we may scale the solution setting $d_s = 0$. Under this assumption all feasible vectors d must be lower bounds for the distances computed by Dijkstra's algorithm. Similarly, we derive lower bounds for the distances for instances of the problem without negative circuits. For instances with a negative directed circuit the problem is infeasible, since such a circuit is in the kernel of B but has a negative scalar product with w.

Now consider the arcs where the inequality becomes an equality. Then any directed s-t-path using only such arcs must be shortest.

What happens if we dualize this program?

$$\min w^\top x$$
$$\text{subject to} \quad B^\top x = (\chi(t) - \chi(s))$$
$$x \geq 0.$$

An interpretation of this program is derived as follows. The variables are indexed by arcs, the arcs emanating from s have a total "flow" $\sum_{(s,v) \in E} x_{s,v}$ of 1. For all vertices different from s and t the sum of the variables of the incoming arcs shall equal the sum of the variables of the outgoing arcs; i.e., the variables shall respect *Kirchhoff's Law of Flow Conservation*, and all the flow shall end in t. Thus we ask for the cheapest possible way to send one unit of flow from s to t. Clearly, this has to be done on a shortest path.

Exits

Shortest path problems are quite fundamental in applications, and algorithmic variants can provide large speed-ups in specific instances or graph classes. For example the sequence alignment problem (Application 4) is central in bioinformatics and comes in many variants aligning very long sequences (whole chromosomes [16]) or multiple shorter sequences (multiple sequence alignment [44]). Mostly, the problem is formulated as dynamic programming (Bellman's 1957 classic paper has been re-published [2]) which only implicitly reflects the (infinite) grid graph presented in the previous section. Algorithmic improvements are obtained by identifying portions of the grid graph through which a shortest path cannot pass through a range of exact or heuristic ideas. This is closely related to the Viterbi-algorithm [43, 34] in Hidden Markov Models which computes the most likely state sequence for an observation sequence, which, for example, is used to recognize speech.

There are much more complicated variants of the shortest path problem in its colloquial interpretation. A shortest path often refers to the fastest way from A to B, not the one with a minimal distance measured in miles. Travel times however are not deterministic but rather stochastic (... it depends on the traffic) and moreover, they are not static but rather dynamic depending on a number of variables (... accident on I-95). The theory of stochastic [19, 3] and dynamic [20, 35] shortest paths reflect those realities.

A generalization of the `FindPathEuclid` algorithm is known as the A* algorithm [23]. We obtain a large speed-up in the Euclidean variant of Dijkstra's algorithm because we can easily ignore vertices unlikely to be on a shortest s-t path, as the Euclidean distance between any vertex v and t provides a lower bound for the distance between v and t in the graph. The existence of the lower bound is crucial for the speed-up, the fact that the graph is Euclidean implies that such an easily computable bound exists, but this is not the only situation in which it does. In the artificial intelligence community the A* algorithm is widely used for planning and game play. Often, the lower bound is replaced by a heuristic estimating the distance between v and t in the graph. There is a large body of literature concerning theory, variants, and applications; see for example [36].

Exercises

Exercise 35 Let $D = (V, A)$ be a directed graph with a weight function $length$: $E \rightarrow \mathbb{Z}$ a on the arcs and an additional cost $c : V \rightarrow \mathbb{Z}$ on the vertices and $s, t \in V$. The total cost $C(P)$ of an s-t-path P is given as

$$\sum_{e \in P} length(e) + \sum_{v \in P} c(v).$$

Give a reduction of this problem to the shortest-path problem.

Exercise 36 Design an algorithm that, given a weighted digraph $D = (V, A)$, $length$: $E \rightarrow \mathbb{Z}$ and $s \in V$, computes an arc a of the digraph such that

$$\min\{dist_{D \backslash a}(s, v) - dist_D(s, v) \mid dist_{D \backslash a}(s, v) - dist_{D \backslash a}(s, v) > 0\}$$

is maximized.

Exercise 37 Let $G = (V, E)$ be a directed acyclic graph (DAG). Is there a linear time, $O(|V| + |E|)$, algorithm for computing a shortest s-t-path? Hint: Exercise 14.

Exercise 38 Recall that we can use various algorithms presented in this chapter to compute shortest paths from a given vertex s to all other vertices in an undirected graph with strictly positive edge weights. The output of such an algorithm is called a *shortest path tree*.

(i) Give an example of a weighted graph and a given vertex s such that no shortest path tree is a minimum spanning tree.
(ii) Prove that any shortest path tree and any minimum spanning tree have at least one edge in common.

Exercise 39 Consider the following *knapsack problem*. Given a list of n items with volumes $c_1, \ldots, c_n$ and utilities $a_1, \ldots, a_n$ and a capacity C of our knapsack. Write a dynamic program to compute a set $I \subseteq \{1, \ldots, n\}$ that maximizes

$$\left\{ \sum_{i \in I} a_i \mid \sum_{i \in I} c_i \leq C \right\}.$$

That is, we want to pack our knapsack maximizing utility and respecting the capacity constraint.

Exercise 40 Consider a set C of n different currencies and a matrix $R = (r_{ij})$ of exchange rates. That is, you receive r_{ij} units of currency i for one unit of currency j. An *arbitrage* is a sequence of currency changes starting and ending with a fixed currency i that generates profit. Write an algorithm that detects possibilities for arbitrage in such a matrix R.

Chapter 6
Maximal Flows

6.1 Introduction

The last chapter ended with a remark that one might consider the shortest path problem as the task to send one unit of "flow" from s to t through a network at minimum cost. In more realistic problems from transportation we usually have to deal with capacity constraints, limiting the amount of flow across arcs. Hence, the basic problem in a capacitated network is to send as much flow as possible between two designated vertices, more precisely from a source vertex to a sink vertex. This notion of a flow in a capacitated network is made more precise in the following.

Definition 16 Let $D = (V, A, cap)$ be a directed network with a capacity function cap: $A \to \mathbb{R}_+$ on the arcs. We call this a capacitated network. An s-t-flow $f : A \to \mathbb{R}_+$ for two designated vertices s, t $\in$ V is a function satisfying

Flow conservation: *the flow out of each vertex $v \in V \setminus \{s, t\}$ equals the flow into that vertex, that is*:

$$\sum_{\substack{w \in V \\ (v,w) \in A}} f(v, w) = \sum_{\substack{w \in V \\ (w,v) \in A}} f(w, v).$$

Capacity constraints: *The flow on each arc $(u, v) \in A$ respects the capacity constraints. That is,*

$$0 \le f(u, v) \le cap(u, v).$$

We define the *net flow from s* to be

$$|f| := \sum_{\substack{w \in V \\ (s,w) \in A}} f(s, w) - \sum_{\substack{w \in V \\ (w,s) \in A}} f(w, s).$$

W. Hochstättler, A. Schliep, *CATBox*, DOI 10.1007/978-3-642-03822-8_6,
© Springer-Verlag Berlin Heidelberg 2010

The flow conservation rules immediately imply

$$\forall v \in V \setminus \{s, t\} : \sum_{\substack{w \in V \\ (v,w) \in A}} f(v, w) - \sum_{\substack{w \in V \\ (w,v) \in A}} f(w, v) = 0.$$

Adding all these equations to the net flow from s yields

$$|f| := \sum_{\substack{w \in V \\ (s,w) \in A}} f(s, w) - \sum_{\substack{w \in V \\ (w,s) \in A}} f(w, s)$$

$$= \sum_{\substack{v \in V \setminus \{t\} \\ (v,w) \in A}} f(v, w) - \sum_{\substack{v \in V \setminus \{t\} \\ (w,v) \in A}} f(w, v)$$

$$= \sum_{\substack{w \in V \\ (w,t) \in A}} f(w, t) - \sum_{\substack{w \in V \\ (t,w) \in A}} f(t, w),$$

thus the *net flow from s* equals the net flow into t. We can state our problem now as:

Problem 8 Let $G = (V, A, cap)$ be a capacitated network, $s, t \in V$. Find an s-t-flow which maximizes the net flow from s.

It will make our discussion a bit easier if we restrict the capacity function to integer values for a while, thus we get:

Problem 9 Let $G = (V, A, cap)$ be a capacitated network with an integer capacity function $cap : A \to \mathbb{Z}_+$ and $s, t \in V$. Find an s-t-flow, maximizing the net flow from s.

6.2 The Algorithm of Ford and Fulkerson

Considering shortest path problems we realized that we might consider paths as flows of unit value. The first algorithm for the maximum flow problem that we will introduce, successively finds paths from s to t. It takes the already existing flow into account and augments it by sending flow along that path. Clearly, a flow reduces the remaining capacity. A priori, it is not clear that we can send flow freely without making any mistakes.

Software Exercise 41 Start the software and open the graph FordFulkerson4. cat. If you point on the edges you can check their capacity in the info line at the bottom of the window. Vertices s and t in these examples are always those with smallest respectively largest number. The first idea might be to send flow on the three edges that make up a path from 1 to 8. The bottlenecks of the capacity of these edges are the first and the last edge with a capacity of 150. Assume we send 150 units of flow over this path. If we just subtract these numbers from the capacities the graph becomes disconnected and there no longer exists a path from s to t. Nevertheless, this flow is not maximal. We can achieve a larger flow by sending 141 units of flow on each of the paths 1,3,4,7,8 and 1,2,5,6,8 and 9 units right through the middle.

This problem of not being able to revert incorrect previous choices is overcome by the idea of the backward arc. When sending flow through the network we add an artificial backward arc for each arc of nonzero flow, indicating that this flow "can be sent back", thus allowing to reduce the flow on the corresponding forward arc. Formally we get

Definition 17 Let $D = (V, A, cap)$ be a capacitated network and $f : A \to \mathbb{R}_+$ an s-t-flow. The residual network $RN(D, f)$ with respect to f is the network $(V, \widetilde{A}, rescap)$. For an arc $a \in A$ we denote by $-a$ a copy of a where $head$ and $tail$ have been interchanged and we write $-A := \{-a \mid a \in A\}$ for all such arcs. The residual capacity we define as

$$rescap\,(a) = \begin{cases} cap\,(a) - f(a) & \text{for } a \in A \\ f(-a) & \text{for } a \in -A \end{cases}$$

and finally $\widetilde{A} := \{\tilde{a} \in A \cup -A \mid rescap\,(a) > 0\}$.

With this definition it is easy to design our first algorithm. We iteratively search for a path P in the residual network and send as much flow as possible through this path. For $P \subseteq A$ let $\chi(P) \in \{0, \pm 1\}^A$ denote the *signed characteristic function* of P, i.e. the vector where we have a 1 in the indices of forward arcs from A in P, a -1 in the A-index if the corresponding arc in $-A$ is used by the P and zeroes elsewhere, that is

$$\chi(P)_a := \begin{cases} 1 & \text{if } a \text{ is a forward arc of } P \\ -1 & \text{if } a \text{ is a backward arc of } P \\ 0 & \text{if } a \text{ is a not an arc of } P \end{cases} \tag{6.1}$$

In the algorithm we search for an s-t path in the residual network, compute its bottleneck delta and update the flow $f = f + \texttt{delta}\,\chi(P)$. The latter is done by recursively backtracking the path in a for loop and computing $\texttt{flow[(u,v)]} = \texttt{flow[(u,v)]} \pm \texttt{delta}$ where the sign is chosen according to whether the current arc is an original arc of the network, a $\texttt{ForwardEdge}$, or an artificial backward arc.

In our implementation we compute the path by a BFS in $\texttt{ShortestPath(s,t)}$ and find a path using as few edges as possible. We will see later on that this is a clever choice, but for the following discussion we allow arbitrary paths.

ALGORITHM FordFulkerson

```
def UpdateFlow(Path):
    delta = MinResCap(Path)
    for (u,v) in Edges(Path):
        if ForwardEdge(u,v):
            flow[(u,v)] = flow[(u,v)] + delta
        else:
            flow[(v,u)] = flow[(v,u)] - delta

s = PickSource()
t = PickSink()

while not maximal:
    Path = ShortestPath(s,t)
    if Path:
        UpdateFlow(Path)
    else:
        maximal = true

ShowCut(s)
```

Software Exercise 42 Continuing our last experiment we recommend you to choose the "Window Layout" with "Two graph windows". If you start the software it will indicate the capacity of the arcs by their thickness in the upper window. If you choose "1" to be the source and "8" to be the sink, the algorithm will stop at the first breakpoint where the s-t-path using 3 edges has been found by BFS. Here the vertices not on the path are displayed in grey, if they have been processed in the BFS, blue if they are visited and green if they have not been put into the queue. If we now "Trace" into the "UpdateFlow"-Procedure, first we find the last edge to be the bottleneck and update the flow backtracking along the path accordingly. Simultaneously with the flow, we update the residual network, i.e. we introduce backward arcs and update the capacity if necessary.

The first and the last arc on the path disappear since their capacity has been completely used. If you point on the forward respectively backward edge of the middle arc you will find that the capacity of the forward arc has been reduced by 150 which is exactly the capacity of the backward arc.

The second path we find uses the backward edge and UpdateFlow "repairs" the mistake we made in the first step. The flow on the edge $(3, 6)$ is reduced by the capacity of the bottleneck $(7, 8)$ on the path, i.e. by 141 and we end with the flow described in our previous discussion.

After the update there is no longer any directed s-t-path in the residual network, which is proven by an s-t-cut (to be defined immediately) in the residual network that contains backward arcs only.

6.3 Max-Flow-Min-Cut

We have a simple argument that this algorithm will terminate as long as the capacities are integers. With every path augmentation the flow is incremented by at least one and there are natural upper bounds on the maximal value of the flow, which are called *cuts*.

Definition 18 Let $D = (V, A)$ be a digraph, $s, t \in V$ and $\emptyset \neq S \subset V$. The cut $[S, V \setminus S]$ *induced by* S consists of those edges that have one end in each of the two non-empty sets of the partition $S, V \setminus S$:

$$[S, V \setminus S] := \{(i, j) \in A \mid \{i, j\} \not\subseteq S \text{ and } \{i, j\} \not\subseteq V \setminus S\}.$$

The *forward edges* $(S, V \setminus S)$ of a cut are defined as

$$(S, V \setminus S) := \{(i, j) \in A \mid i \in S, j \in V \setminus S\}.$$

The *backward arcs* of a cut $[S, V \setminus S]$ are denoted as $(V \setminus S, S)$. A cut $[S, V \setminus S]$ is an s-t-cut, if $s \in S$ and $t \notin S$.

If $cap : A \rightarrow \mathbb{R}_+$ is a capacity function, the capacity $cap[S]$ of an s-t-cut is defined as $cap[S] := cap(S, V \setminus S) := \sum_{a \in (S, V \setminus S)} cap(a)$, the total capacity of the forward edges.

The capacity of any cut is an upper bound on the flow value:

Lemma 12 (weak duality) *Let $D = (V, A, cap)$ be a capacitated network, f a flow and $[S, V \setminus S]$ an s-t-cut. Then $|f| \leq cap[S]$.*

At the end of the proof we use the abbreviation $f(\widetilde{E}) := \sum_{e \in \widetilde{E}} f(e)$ if $\widetilde{E} \subseteq E$ is a set of edges (c.f. Theorem 9).

Proof

$$
\begin{aligned}
|f| &= \sum_{\substack{w \in V \\ (s,w) \in A}} f(s, w) - \sum_{\substack{w \in V \\ (w,s) \in A}} f(w, s) \\
&= \sum_{\substack{w \in V \\ (s,w) \in A}} f(s, w) - \sum_{\substack{w \in V \\ (w,s) \in A}} f(w, s) \\
&\quad + \sum_{v \in S \setminus s} \left(\sum_{\substack{w \in V \\ (v,w) \in A}} f(v, w) - \sum_{\substack{w \in V \\ (w,v) \in A}} f(w, v) \right) \\
&= \sum_{v \in S} \left(\sum_{\substack{w \in V \\ (v,w) \in A}} f(v, w) - \sum_{\substack{w \in V \\ (w,v) \in A}} f(w, v) \right) \\
&= f((S, V \setminus S)) - f((V \setminus S, S)) \\
&\leq f((S, V \setminus S)) \\
&\leq cap[S].
\end{aligned}
$$

$\square$

This proves finiteness of the above procedure if the capacities are integers. Furthermore, when the algorithm terminates source s and destination t are separated in the residual network $RN(D, f)$. Thus if we set

$$S^* := \{v \in V \mid \text{ there is some } s-v\text{-path in } RN(D, f)\},$$

then $[S^*, V \setminus S^*]$ is an s-t-cut. We compute $|f| = f((S^*, V \setminus S^*)) - f((V \setminus S^*, S^*))$. Since there is no edge emanating from S^* in $RN(D, f)$ there cannot be any nonzero flow on backward edges of the cut and $f((V \setminus S^*, S^*)) = 0$. Thus $[S^*, V^*]$ is a cut whose capacity equals the value of the flow. Lemma 12 now implies.

Theorem 15 (Max-flow-Min-cut) *Let $D = (V, A, cap)$ be a capacitated network and $s, t \in V$. Then*

$$\max_{f \text{ is } s\text{-}t\text{-flow}} |f| = \min_{[S, V \setminus S] \text{ is } s\text{-}t\text{-cut}} cap[S].$$

If, furthermore, the capacities are integer then a maximal flow can be chosen to be integer on all edges.

Exercise 43 Let $G(V, E)$ be a graph and $s, t \in V$ two vertices.

(i) Prove that the number of pairwise edge disjoint s-t-paths equals the minimal number of edges that must be removed in order to destroy all s-t-paths.
(ii) Prove that the number of pairwise internally vertex disjoint s-t-paths equals the minimal number of vertices that must be removed in order to destroy all s-t-paths.

6.4 Pathologies

In this section we will learn that the augmenting path algorithm with a clumsy implementation can have an exponential running time behavior. Before we do so, we will present an example with irrational data, where the algorithm not only does not terminate but also converges towards a non-optimal value.

Example 4 Consider the network in Fig. 6.1. The most important edges are the three innermost edges, the upper and the lower run from left to right with capacities of 1 respectively $\sqrt{2}$. The edge in the middle runs backwards and has an infinite capacity as well as all other edges except of the direct arc (s, t) which has a capacity of 1. The latter edge—as well as several backward arcs—will be omitted in the following pictures since it is not used until the very end. After the first four indicated augmenting steps we get into a situation where the capacities of all arcs—except for the direct arc—are a scalar multiple of the data in the beginning. Clearly an infinite repetition of these steps will thus converge towards a flow of value $1 + \sqrt{2}$, while a flow of value $2 + \sqrt{2}$ exists.

Even with integer data a bad implementation of the above procedure may result in an exponential run time behavior.

Example 5 Consider the network depicted in Fig. 6.1. If each augmentation uses the "bottleneck edge", clearly we will need 2^{k+1} augmentations in order to compute a maximal flow. The size of the data though is dominated by the coding length of the numbers and is $O(k)$. Thus the number of augmentations is exponential in the size of the data.

Software Exercise 44 The interested reader may ask why we did not animate these examples with our software. While this is quite obvious for the example with irrational data, you may try it yourself. Load the file `FordFulkersonWC.cat` and run `FordFulkerson.alg` on it. Isn't it disappointing, it terminates in two

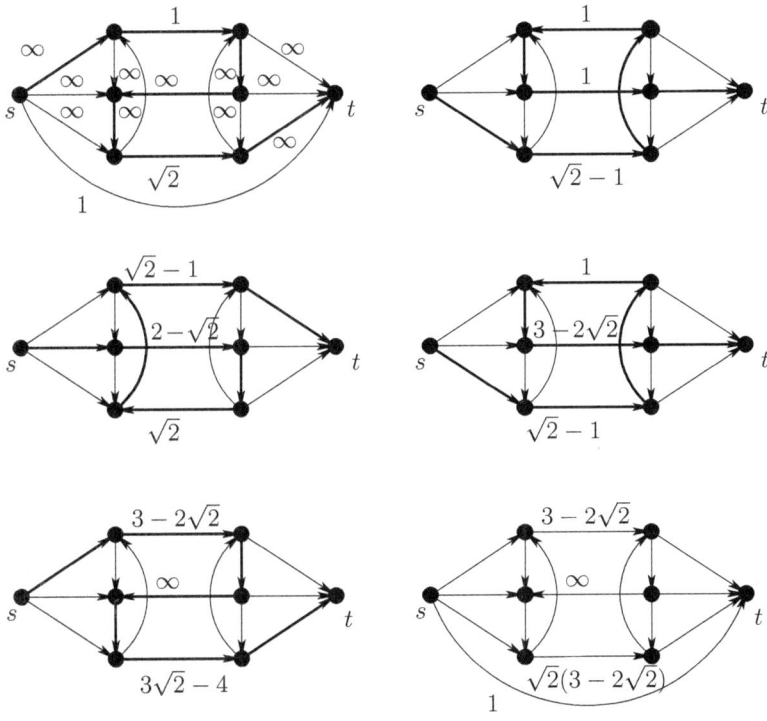

Fig. 6.1 Pathological irrational

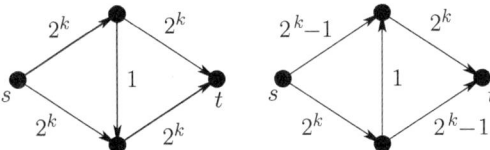

Fig. 6.2 A smallest worst case example

iterations. We consider it difficult to find some reasonable implementation that would chose the augmenting paths as in the above examples. This does not mean that there is no problem. The difficulties may occur in larger examples less obviously and hidden somewhere, but actually not with the implementation in our package. The reason will become clear in the next section.

6.5 Edmonds-Karp Implementation

The bad example in the last section suggests two possible approaches to designing a good implementation. Either by shortest paths, i.e., paths using as few edges as possible, or by paths of maximal capacity. The latter can be implemented, but is "unlikely to be efficient" [1].

The former is quite natural and it is the way the augmenting paths are chosen in our implementation of Ford and Fulkerson's algorithm. We run a BFS in the residual network, starting at s until we reach t, i.e. we search for a path from s to t that uses as few edges as possible. The advantage of this implementation is, that we have additional control over the flow of the algorithm as the length of the shortest paths from any vertex v to t must be monotonically increasing:

Lemma 13 *Let $D = (V, A, cap)$ be a capacitated network (cap not necessarily rational), f a flow, P a directed s-t-path in $RN(D, f)$ using as few edges as possible and $\tilde{f}$ the flow that arises from f by augmentation along P, $\tilde{f} = f + \Delta \chi(P)$. Let $dist(v)$ denote the distance from v to t in $RN(D, f)$, with respect to unit weights, that is the number of edges, and $\widetilde{dist}(v)$ the same in v in $RN(D, \tilde{f})$, then*

$$\forall v \in V : \widetilde{dist}(v) \geq dist(v).$$

Proof $RN(D, f)$ and $RN(D, \tilde{f})$ differ only along P. Edges of that path which had the residual capacity of the bottleneck of this path will disappear. Possibly we will be facing newly introduced backward arcs for edges that had no flow in f. Let (i, j) be such an edge. Since (i, j) belongs to a shortest s-t-path in $RN(D, f)$ we necessarily have $dist(i) = dist(j) + 1$.

We now show the assertion by induction on $\widetilde{dist}(v)$. If $\widetilde{dist}(v) = 0$ then $v = t$ and there is nothing to prove. Thus, let $\widetilde{dist}(v) = k$ and $v = v_0, \ldots, v_k = t$ be a shortest v-t-path in $RN(D, \tilde{f})$. If edge (v, v_1) is an edge in $RN(D, f)$ as well, we have $dist(v) \leq dist(v_1) + 1 \leq \widetilde{dist}(v_1) + 1 = \widetilde{dist}(v)$ by inductive assumption. Otherwise, (v_1, v) was on a shortest s-t-path in $RN(D, f)$ and we get $dist(v) = dist(v_1) - 1 \leq \widetilde{dist}(v_1) - 1 = \widetilde{dist}(v) - 2$.   □

As an immediate corollary we get

Corollary 3 *Each edge is saturated at most $\dfrac{|V|}{2}$ times in each direction, i.e. the flow on that edge is set to maximal capacity, or to zero respectively.*

Proof   Whenever an edge (i, j) is saturated then $dist(i) = dist(j) + 1$. In order to get saturated again in the same direction, it has to occur in a shortest path in the other direction, first. The distances $\widetilde{dist}$, valid at that moment, satisfy $\widetilde{dist}(j) = \widetilde{dist}(i) + 1 \geq dist(i) + 1 = dist(j) + 2$. As the distances are bounded by $|V|$, the claim follows.   □

Theorem 16 *The implementation of Edmonds and Karp computes a maximal flow in $O(|A|^2|V|)$ steps (even if the data is irrational).*

Proof   According to the last Corollary we need at most $O(|A||V|)$ augmentations. The bound follows, since we can find a shortest path in $RN(D, f)$ and update the residual network in $O(|A|)$.   □

Remark 9 A complete recomputation of the distances from scratch in each augmentation is not efficient. By a clever update strategy on the distance it is possible to achieve an overall complexity of $O(|A||V|^2)$ (see again [1] for details).

6.6 Max-flow-Min-cut as Linear Programming Duality

The definition of our network flow problem has a strong algebraic flavor and it is already quite close to its linear program, which is as follows:

$$\max (-d(s))^\top x$$
$$\text{subject to} \quad (-B)x = 0$$
$$x \leq w$$
$$x \geq 0,$$

where $d(s) \in \{0, \pm 1\}^A$ is the directed incidence vector of s, that is

$$d(s)_e = \begin{cases} -1 & \text{if } e = (s, i) \\ 1 & \text{if } e = (i, s) \\ 0 & \text{if } e = (i, j), \ i \neq s \neq j \end{cases}.$$

B is the vertex-arc incidence matrix, where the rows indexed by s and t have been deleted.

Going over to the dual we derive the following program:

$$\min \quad u^\top w$$
$$\text{subject to } u - p^\top B \geq -d(s)$$
$$u \geq 0,$$

where u are edge variables and p vertex variables. We interpret the value of a p-variable p_v as the *potential* of vertex v. How does such a dually feasible potential change if we follow its values along an s-t-path $s = v_0, \ldots, v_k = t$ with edge set U? The inequalities involving variables corresponding to edges from U are

$$-p_{v_1} + u_{s,v_1} \geq 1,$$
$$p_{v_i} - p_{v_{i+1}} + u_{i,i+1} \geq 0,$$
$$p_{v_{k-1}} + u_{v_{k-1},t} \geq 0.$$

Adding these inequalities yields $\sum_{e \in U} u_e \geq 1$.

If we introduce a potential $p_s = -1$ for s and $p_t = 0$ for t, then all of the above inequalities look alike. Each time we have an increase of Δ in potential from v_i to v_{i+1}, then this increase has to be balanced by the edge variable $u_{i,i+1} \geq \Delta$. Since our objective is to minimize $\sum_{e \in E} w_e u_e$ and there has to be a total increase in potential of 1 along each s-t-path, this is achieved at minimum cost only if we realize it across a minimum cut:

Lemma 14 *Let $D = (V, A, cap)$ be a capacitated network and $s, t \in V$, such that every vertex $v \in V$ lies on some directed s-t-path. Let*

$$MC := \{u \in \mathbb{R}^A \mid \exists p \in \mathbb{R}^V : (p, u) \text{ is feasible for the dual program}\}.$$

Then $u \in MC$ if and only if there exist (directed) s-t-cuts $C_i = (S_i, V \setminus S_i)$ and coefficients $\lambda_i \in [0, 1]$, $i = 1, \ldots, k$ such that $\sum_{i=1}^{k} \lambda_i = 1$ and $u \geq \sum_{i=1}^{k} \lambda_i \chi(C_i)$. The latter means that MC is the sum of the convex hull of all s-t-cuts plus the cone of the positive orthant:

$$MC = conv\{\chi(C) \mid C \text{ st } s-t-cut\} + \mathbb{R}_+^A.$$

Proof Let $C = (S, V \setminus S)$ be a directed s-t-cut. Setting

$$p_v = \begin{cases} -1 & \text{if } v \in S \\ 0 & \text{if } v \notin S, \end{cases}$$

we see that $(p, \chi(C))$ is feasible for the dual program. Hence, by Lemma 7 $conv\{\chi(C) \mid C \text{ is } s-t-cut\} \subseteq MC$. Clearly, we can add any positive vector without leaving MC and thus one inclusion follows.

For the other inclusion let (p, u) be feasible for the dual program. We proceed by induction on the number of non-zeroes in u. Let S denote the set of vertices that are reachable from s on a directed path that uses only edges a satisfying $u_a = 0$. Then $t \notin S$ and thus $C_1 = (S, V \setminus S)$ is an s-t-cut. Let $\lambda_1 := \min_{a \in (S, V \setminus S)} u_a$. If $\lambda_1 \geq 1$, the claim follows and founds our induction. Otherwise, set $\tilde{u} = \frac{1}{1-\lambda_1}(u - \lambda_1 \chi(C_1))$ and furthermore

$$\tilde{p}_v = \begin{cases} -1 & \text{if } v \in S \\ \frac{1}{1-\lambda_1} p_v & \text{if } v \notin S. \end{cases}$$

We claim that $(\tilde{p}, \tilde{u})$ is feasible for the dual program. Clearly, $\tilde{u}$ has at least one nonzero less than u. Inequalities that correspond to non-cut edges from $A \setminus C$ are immediately seen to be satisfied. Consider an edge $a = (i, j) \in C$. Note, that by definition of S all vertices $i \in S$ are reached on paths using only arcs $\tilde{i}$ where $u_{\tilde{a}} = 0$ and hence, necessarily, $p_i \leq -1$ must hold. Using this we compute

$$
\begin{aligned}
\tilde{p}_i - \tilde{p}_j + \tilde{u}_a &= -1 - \frac{1}{1 - \lambda_1} p_j + \frac{1}{1 - \lambda_1}(u_a - \lambda_1) \\
&= \frac{1}{1 - \lambda_1}(\lambda_1 - 1 - p_j + u_a - \lambda_1) \\
&\geq \frac{1}{1 - \lambda_1}(p_i - p_j + u_a) \\
&\geq 0.
\end{aligned}
$$

By inductive assumption there exist directed cuts $C_2, \ldots, C_k$ and coefficients $\mu_2, \ldots \mu_k$ such that $\tilde{u} \geq \sum_{i=2}^{k} \mu_i \chi(C_i)$ and $\sum_{i=2}^{k} \mu_i = 1$. Putting $\lambda_i = (1 - \lambda_1)\mu_i$ for $i = 2, \ldots, k$ yields the desired combination. □

Summarizing, the optimal solution of the dual LP is the incidence vector of a minimal s-t-cut, or at least a convex combination of minimal cuts, and therefore MaxFlow-MinCut is an example of LP-duality.

6.7 Preflow Push

In the algorithm of Ford and Fulkerson we have a feasible flow at any time and thus a feasible solution of the linear program. The minimal cut at the end shows up at sudden. We could try to approach this problem from a dual point of view. A disadvantage of the above method is that the computation of an augmenting path always takes $O(|V|)$ steps, because we compute it from scratch. How bad this may be is best visualized by an example.

Software Exercise 45 Run our Edmonds Karp implementation of the Ford Fulkerson algorithm on the graph EdmondsKarpWC.cat. It consists of a path of large capacity and a mesh of several paths of length two and capacity one. Each time we increase the flow by a unit we have to recompute the unique path again and again.

The last example illustrates one basic idea of the preflow push algorithms due to Goldberg and Tarjan: Push as much flow onward as possible. Unfortunately, it is in general not clear a priori what "onward" means. In order to overcome this, we interpret it the following way:

- The (primal) balancing conditions are violated at several vertices, we have an excess of flow into a vertex and we try to fix this by pushing flow, if possible towards the sink.
- Simultaneously we keep a "reference cut" of non-increasing capacity that is used completely by the flow.

The procedure terminates, when the "preflow" becomes a flow, i.e. primally feasible. The main problem is to determine the direction into which the flow shall be pushed. The idea to solve this problem stems from the following physical model. Consider a network of pipelines of the given capacity, where the nodes of the network may be lifted. Naturally, flow goes downhill. Our intention is to send as much flow as possible from s to the sink t. The sink is fixed at ground zero and s at a height of $|V|$. In a preprocessing step we determine for each vertex $v \in V \setminus \{s, t\}$ its minimum height, namely, $dist(v, t)$, the length of a shortest v-t-path. This is the minimal descent that is required for any flow passing v to have a chance to arrive at t. If, now, we let the flow drip from s, due to capacity constraints, an excess might get stuck at some points. These are lifted in order to diminish the excess by allowing some flow back towards the source. Note, that this way, some flow may now go uphill. We will see that, since the height of s and t is fixed, this procedure is finite and in the end some of the initially sent flow will arrive at t and the remaining be returned to s.

Putting this idea into an algebraic framework we define:

Definition 19 Let $D = (V, A, cap)$ be a capacitated network and $s, t \in V$. An s-t-preflow $f : A \to \mathbb{R}_+$ is a function satisfying

preflow condition: For all vertices $v \in V \setminus \{s\}$ the preflow out of the vertex does not exceed the preflow into the vertex: $\sum_{(v,w) \in A} f(v, w) \leq \sum_{(w,v) \in A} f(w, v)$,
capacity constraints: For all arcs $(u, v) \in A$: $0 \leq f(u, v) \leq cap(u, v)$.

Let f be a preflow. The residual network $RN(D, f)$ is defined as in the case of a flow. If $v \in V \setminus \{s, t\}$ then the excess $ex(f, v)$ of f in v is defined as $\sum_{(w,v) \in A} f(w, v) - \sum_{(v,w) \in A} f(v, w)$. A vertex with positive excess is called *active*.

We implement the idea described above as follows. The height of vertex v is stored in the variable pot[v], which we call the potential of v. This potential should not be confused with the dual variables of the linear program. Now, as long as some vertex v has a positive excess(v), we try to find a vertex u in Neighborhood(v) with pot[u] < pot[v] and push as much flow as possible, namely the minimum of the residual capacity res((v,u)) and the excess of the vertex, along (v,u). If such a vertex does not exist, we enforce its existence for the next iteration by updating pot[v] = minResNeighborPot(v)+1 the potential of v to one more than the minimal potential of a neighbor reachable through a directed arc with a non-vanishing residual capacity.

ALGORITHM PreflowPush

```
InitPotential(s,t)
for v in G.Neighborhood(s):
    flow[(s,v)] = cap((s,v))

8  feasible = False
   while not feasible:
       pushed = False
       v = FindExcessVertex()
       if not v:
13         feasible = True
       else:
           for u in Neighborhood(v):
               if pot[v] > pot[u]:
                   delta = Min(res((v,u)),excess(v))
18                 if ForwardEdge(v,u):
                       flow[(v,u)] = flow[(v,u)] + delta
                   else:
                       flow[(u,v)] = flow[(u,v)] - delta
                   pushed = True
23                 break
           if not pushed:
               pot[v] = minResNeighborPot(v)+1

ShowCut(s)
```

The line `pot[v]=minResNeighborPot(v)+1` is feasible which can be seen as follows. If $v \in V \setminus \{s, t\}$ satisfies $ex(f, v) > 0$ and hence

$$\sum_{(u,v)\in A} f(u, v) > \sum_{(v,u)\in A} f(v, u),$$

then it must have some neighbor w such that $f(w, v) > 0$ and as a consequence w is a neighbor in the residual network.

It is not clear a priori, that the algorithm is finite and we have to work on that. Interestingly, we will not directly prove that the flow must become feasible, but prove instead that the labels `pot` are bounded. For that purpose, we first show that the vertex labels can be considered as lower bounds for the distances in the residual network. This will be an immediate consequence of the following Proposition:

Proposition 5 *During execution of the while-loop at any time we have* `pot[i]` $\le$ `pot[j]` $+ 1$ *for all edges* $(i, j) \in RN(D, f)$.

Proof In the beginning this is guaranteed by the preprocessing. Whenever additional flow is sent on an edge (i, j), by inductive assumption we have `pot[i]` $\le$ `pot[j]` $+ 1$. Since we send flow along that arc on the other hand we necessarily have `pot[i]` $>$ `pot[j]`, thus `pot[i]` $=$ `pot[j]` $+ 1$. In particular, if an arc (j, i) is newly introduced in the residual network it satisfies `pot[j]` $=$ `pot[i]` $- 1 \le$ `pot[i]` $+ 1$. If a label is increased, the assertion directly follows from the inductive assumption. □

Now, we can bound the minimum length of a path by the difference in potential

Corollary 4 *If* `pot[v]` $= k$ *and* `pot[w]` $= l > k$, *then any directed w-v-path in the residual network* $RN(D, f)$ *uses at least* $l - k$ *edges.*

Next, we consider the situation that for some vertex v there no longer exists a directed v-t-path, but it has a positive excess. What do we do then? We increase its label, but the backward arcs of the flow, that accumulates in v, bound the labels, since there always exists an s-v-path in this situation:

Proposition 6 *If* $v \in V \setminus \{s, t\}$ *and* $ex(f, v) > 0$ *then there exists a directed* v-s-path *in* $RN(D, f)$.

Proof Let S denote the set of vertices v that are connected to s by a directed v-s-path. Clearly $s \in S$. By definition of $RN(D, f)$ we cannot have a non-zero flow on any edge $e \in (S, V \setminus S)$ leaving S. The preflow condition for $V \setminus S$ implies

$$0 \leq \sum_{v \in V \setminus S} ex(f, v) = \sum_{v \in V \setminus S} \left(\sum_{(u,v) \in A} f(u, v) - \sum_{(v,w) \in A} f(v, w) \right)$$

$$= \sum_{e \in (S, V \setminus S)} f(e) - \sum_{e \in (V \setminus S, S)} f(e).$$

As there is no flow on any arc leaving S we get

$$0 = \sum_{e \in (S, V \setminus S)} f(e) \geq \sum_{e \in (V \setminus S, S)} f(e) \geq 0.$$

Hence all there is no flow on the arcs of $(V \setminus S, S)$ implying

$$\sum_{v \in V \setminus S} ex(f, v) = 0.$$

As the preflow condition guarantees a non-negative excess, we must have $ex(f, v) = 0$ for all $v \in V \setminus S$. $\qquad\square$

Since the label of s is fixed to $|V|$, we get the following Corollary, which implies finiteness of the algorithm.

Corollary 5 *Throughout the algorithm and for all* $v \in V$ *we have:* pot[v] $\leq 2|V| - 1$.

Proof The label of a vertex is increased only if it has a positive excess. But then there exists a directed path to s in the residual network and the claim follows from Corollary 4. $\qquad\square$

As now the finiteness of the algorithm has become clear, we examine its time complexity. In each iteration of the while loop we either increase the potential of a vertex or we perform a push on some edge. By Corollary 5 the number of relabel operations is of order $O(|V|^2)$. The same considerations we made for the Edmonds-Karp implementation of the Ford-Fulkerson algorithm in Corollary 3 yields an upper bound of $O(|V||A|)$ on the number of saturating pushes. In order to bound the number of non-saturating pushes we use the trick of a "potential function".

Lemma 15 *The number of non-saturating pushes in a run of the preflow push algorithm is $O(|V|^2|A|)$.*

Proof   We consider the sum of the labels of the active vertices

$$\Phi := \sum_{i \text{ is active}} \texttt{pot[i]}$$

as a *potential function*. By Corollary 5 this number is bounded by $2|V|^2$. When the algorithm terminates we have $\Phi = 0$. Consider a non-saturating push along (i, j). Afterwards i is no longer active but j is active. As $\texttt{pot[j]} = \texttt{pot[i]} - 1$ this implies:

A non-saturating push decrements Φ by at least 1.

The claim now follows if we can show that the total increase of Φ is of order $O(|V|^2|A|)$. We consider the two possible cases of an increase of Φ.

Case 1: A saturating push is performed. This may result in a new active vertex. Therefore the potential function may grow, but at most by $2|V| - 1$. As there are $O(|V||A|)$ saturating pushes we can bound the total increase of Φ caused by saturating pushes by $O(|V|^2|A|)$.

Case 2: A vertex is relabeled. Since the labels never decrease and are bounded by $2|V| - 1$, the total increase caused by relabeling operations is of order $O(|V|^2)$.

$\square$

Summarizing we have the following bound on the running time of our first version of the preflow push algorithm.

Theorem 17 *The running time of the preflow push algorithm is of order $O(|V|^2|A|)$.*

You may have realized that we did not make any comment on the correctness of the algorithm, yet. All we know by now is that it computes a flow, since there are no active vertices left at the time of termination. We will conclude correctness directly from the existence of a cut of the same size using linear programming duality.

6.8 Preflow Push Considered as a Dual Algorithm

As seen above the initial preflow is infeasible. We will show that it is optimal as soon as it becomes feasible. This is done by providing a saturated cut for the preflow at each stage. This saturated cut is *complementary* to the preflow as its variables are zero if the corresponding inequalities of the capacity constraints are strict; i.e., we always have $(f_a - cap_a)u_a = 0$. Furthermore, the potentials as defined in the proof of Lemma 14 guarantee that $(p_i - p_j + u_{ij})f_{ij} = 0$. To see this, note that the left coefficient in that product is nonzero only on backward arcs of the saturated cut which will have a zero flow.

Recall from Theorem 5 that in general a complementary pair of primally and dually feasible solutions is optimal for both programs.

Recall the following equation about "conservation of mass" that holds for any function f on the arcs if $[S, V \setminus S]$ is an s-t-cut:

$$f((S, V \setminus S)) - f((V \setminus S, S)) = \sum_{v \in V \setminus S} ex(f, v). \qquad (6.2)$$

For our dual solution we define the cuts iteratively. First, we set $S_0 = \{s\}$. Thus, in the beginning of the while loop of the algorithm PreflowPush on page 81 $(S_0, V \setminus S_0)$ is a saturated s-t-cut. Now, if $(S_i, V \setminus S_i)$ is a saturated s-t-cut for some steps in the algorithm we define S_{i+1} when it becomes necessary—because $(S_i, V \setminus S_i)$ is no longer saturated—as follows. As soon as some edge $(u, v_0) \in (S_i, V \setminus S_i)$ becomes an edge of the residual network $RN(D, f)$ we set

$$S_{i+1} := S_i \cup \{v \in V \mid \text{there exists a directed } v_0\text{–}v \text{ path in } RN(D, f)\}.$$

Then the following holds

Lemma 16 S_i is an s-t-cut for all i and

  (i) $S_i \subseteq S_{i+1}$,
 (ii) $cap[S_i] = \sum_{v \in V \setminus S_i} ex(f, v)$,
(iii) $cap[S_i] \geq cap[S_{i+1}]$.

Proof In order to show that $t \notin S_i$ it suffices to show that for all i and all $v \in S_i$ at any time pot[v] $> |V \setminus S_i|$, which implies the assertion as pot[t] $= 0$. By definition this holds for S_0. We proceed by induction on i. Let $u = v_0, v = v_k$ and $v_0 v_1 \ldots v_k$ be a directed path in $RN(D, f)$, where $u \in S_i$ and $\{v_1, \ldots, v_k\} \cap S_i = \emptyset$. If pot[v] $\geq$ pot[u] we are done. Otherwise by Corollary 4, we have pot[u] $-$ pot[v] $\leq k$ and thus pot[v] $\geq$ pot[u] $- k > |V \setminus S_i| - k \geq |V \setminus S_{i+1}|$. Now, assertion 1 is clear by definition.

As by construction, when S_{i+1} is defined, all forward edges of $[S_{i+1}, V \setminus S_{i+1}]$ are saturated and all backward edges have zero flow, the second assertion follows from (6.2). Since the excesses are nonnegative and the sets S_i grow monotonically the third assertion follows from the second. □

Now, we get the correctness of the algorithm as a corollary.

Corollary 6 *The preflow push algorithm computes a maximal flow.*

Proof As mentioned above the preflow has become a flow when the algorithm terminates, i.e. t is the only active node and the net flow into t equals the value of the cut S_i. □

6.9 FIFO-Implementation

We can get an improved bound on the running time by a clever implementation of the choice of active vertices. Furthermore, it seems quite natural to work on an active vertex until it is no longer active or has to be relabeled. We say that we *examine a node*. If we have to relabel a node we append it to the end of the queue.

The choice of the vertex can be done by maximal label or maximal excess to achieve the best bounds of $O(\sqrt{|A|}|V|^2)$ respectively $O(|V||A| + |V|^2 \log(U))$, where U denotes an upper bound on the (integer) capacities (see again [1] for details). For a recent overview on implementations and their complexities we recommend [38].

Here we will discuss the simpler alternative of a First-In-First-Out queue that we already met several times during our course. This means that we always choose the oldest active node, which is at the beginning of the queue. A vertex that becomes active is added to the end of the queue.

The running time is analyzed as follows: We provide a time stamp for each active node. The vertices v that become active in the initialization receive a time stamp of $t(v) = 1$. If vertex v becomes active because edge (u, v) is pushed it receives the time stamp $t(v) = t(u) + 1$. Furthermore the time stamp is incremented if the vertex is relabeled. We will show that using this rule $t \leq 4|V|^2 + |V|$. Recall that the number of non-saturating pushes dominated the running time of our naive implementation and that while examining a node we perform at most one non-saturating push. Hence, the above inequality improves the result of our running time analysis in Sect. 6.7 to $O(|V|^3)$.

As potential function for our analysis here we consider

$$\Phi = \max \{0, \{\texttt{pot[i]} \mid i \text{ is active}\}\}.$$

By Corollary 5, the potential of each vertex is bounded by $2|V| - 1$ and hence the total increase the of the potential function caused by relabel operations is bounded by $|V|(2|V| - 1) \leq 2|V|^2$.

If during the examination of vertices with time stamp k, we call this a *round*, no vertex is relabeled, then the excess of all active vertices has been passed to vertices with smaller label and Φ has decreased by at least 1. Otherwise, a vertex v has been relabeled. In this case, the potential function may increase by at most $\texttt{pot[v]} < 2|V| - 1$. Since the total increase of the sum of all labels (active or non-active) caused by relabel operations is bounded by $2|V|^2$ this can happen in at most $2|V|^2$ rounds and lead to an increase of Φ of at most $2|V|^2$.

As $\Phi < |V|$ in the beginning and since in all other rounds without a relabeling operation we decrease the potential by at least one, there can be at most $|V| + 2|V|^2$ such rounds. We conclude that our algorithms terminates after at most $4|V|^2 + |V|$ rounds and hence all time stamps have a value of at most $4|V|^2 + |V|$. Since no vertex will receive the same time stamp twice and while examining a node we perform at most one non-saturating push the number of non-saturating pushes is bounded by $O(|V|^3)$. Using the analysis of the last section we get the following theorem:

Theorem 18 *The FIFO-implementation of the preflow push computes a maximal flow in* $O(|V|^3)$.

A final remark on "real implementation" of the preflow push.

Software Exercise 46 For an efficient implementation of the preflow push it is mandatory to avoid the clumsy explicit computation of the way that the excess of the preflow is sent back to s. How costly this may be is illustrated by running the algorithm on the example graph `PreflowPushWC.cat`.

One possibility to overcome this problem is to heuristically search for small saturated cuts. The s-side of this cut can be neglected in the following computations.

Exits

The dual pair of problems Max-Flow and Min-Cut is tractable, since in the case of flows we consider only one kind of flow or good that is distributed through the network and two vertices which have to be separated. If we have several goods and pairs of origin and destination, where we want to distinguish the flow we get a "Multiflow"-Problem which is tractable only in a few special cases (see [38]). If we want to separate more than two terminals we have a multiterminal-cut problem which is intractable already for three terminals in general networks [13].

One can generalize the notion of a cut from our definition by dropping the sources and sinks and rather ask about finding minimum cuts in a graph, minimum edge sets which lead to 2 or k connected components when removed. For fixed k, Goldschmidt and Hochbaum [22] give a polynomial time algorithm for this problem, which arises for example in the context of analyzing data by clustering, that is identifying groups of similar objects. In this context the vertices represent objects—e.g., web pages, patients, or pixels in an image—and the edge weight corresponds to the similarity—shared words, common genetic variations, like color and intensities—between objects. Sometimes G is then called a similarity graph and the connected components in the graph without the cut are the clusters. Hence, the k-way cut problem can be viewed as identifying a minimal set of similarities which we ignore to arrive at k clusters. More involved cost functions for cuts, normalized and weighted [8], have been introduced to compensate for artifacts such as unbalanced cluster sizes. Algorithmically, spectral algorithms which rely on the eigenvectors and eigenvalues of the Laplacian of the adjacency matrix, are the state of the art in clustering in similarity graphs [27].

Exercises

Exercise 47 Let $D = (V, A, cap)$ be a capacitated directed network and $S, T \subseteq V$ two disjoint vertex sets. An $S - T$-flow is a function that satisfies flow conservation for all vertices in $V \setminus (S \cup T)$. We define the *value* of the flow

$$|f| := \sum_{s \in S} \left(\sum_{(s,w) \in A} f(s, w) - \sum_{(w,s) \in A} f(w, s) \right)$$

$$= \sum_{t \in T} \left(\sum_{(w,t) \in A} f(w, t) - \sum_{(t,w) \in A} f(t, w) \right).$$

Give a reduction from the problem to find a maximal $S - T$-flow to the task of maximizing a flow from a single source to a single sink.

Exercise 48 Consider a directed network $D(V, A, cap, low)$ with upper and lower bounds on the flow of an arc. An s-t-flow is feasible if it respect these bounds at all arcs and Kirchhoff's law at all vertices different from s and t.

(i) Give a reduction of the problem of the existence of a feasible flow to an ordinary MaxFlow-Problem.
(ii) Modify the algorithm of Ford and Fulkerson in order to compute a maximum flow in a network with upper and lower bounds.

Exercise 49 Write an algorithm to compute an s-t-path of maximum residual capacity in a residual network and analyze its complexity.

Exercise 50 The following is an abstract model of a reload problem where we have to decide to make a delivery directly or via one of two *hubs*. Given a graph $G = (V, E)$, a nonnegative integer weight function $w_0 : E \to \mathbb{Z}^+$ on the edges and two nonnegative integer weight functions on the vertices $w_1, w_2 : V \to \mathbb{Z}^+$. We will say $V' \subseteq V$ *satisfies* an edge $e = (u, v)$ if $\{u, v\} \subseteq V'$. The objective is to find a set of edges $F \subseteq E$ and 2 subsets of the vertices $V_1, V_2 \subseteq V$ such that for all $e = (u, v) \in E$ either $e \in F$ or V_1 or V_2 satisfies e, and

$$\sum_{e \in F} w_0(e) + \sum_{v \in V_1} w_1(v) + \sum_{v \in V_1} w_1(v)$$

is minimized. Model this problem as a MinCut-Problem in a capacitated network.

Chapter 7
Minimum-Cost Flows

7.1 Introduction

Recall application 5 (page 4), modeled by the network in Fig. 1.2. If we consider the edge costs as distances we can compute the optimal production plan for period j as a shortest path. If we have capacities on the edges as well, the problem becomes a combination of a shortest path and a flow problem called a min-cost-flow-problem, which is the kind of problem we will discuss in this chapter. Compared to its simplest form, namely when we interpret the shortest path problem as the task to send one unit of flow from s to t at minimum cost, here, the problem becomes more complex. We may have an arbitrary number of supply and demand vertices, we call them *sources* and *sinks*, respectively. If a vertex $v \in V$ is a source then $b_v > 0$ denotes its *supply*, if v is a sink $b_v < 0$ its *demand*. Supply and demand shall be balanced, i.e. $\sum_{v \in V} b_v = 0$. We have capacities *cap* and costs c on the arcs and the task is to find a flow of minimum cost that transports the supply to the demanding vertices.

We assume that all data is integer and that the digraph has no directed circuit of negative cost and infinite capacity. Thus, if B is the vertex-edge-incidence matrix of the digraph we can formulate the following problem:

Problem 10 Let $D = (V, A, cap)$ be a capacitated network, $cap \in \mathbb{Z}_+^A$, $c : A \to \mathbb{Z}$ a cost function on the arcs and $b : V \to \mathbb{Z}$ a demand function such that $\sum_{v \in V} b(v) = 0$. The *minimum-cost flow problem* then is given as

$$
\begin{aligned}
\min \; & c^\top f \\
\text{subject} \quad & Bf = b \\
\text{to} \quad & f \leq cap \\
& f \geq 0,
\end{aligned}
\tag{7.1}
$$

where $f \in \mathbb{R}^A$ denotes the flow on the arcs. We will start with the development of methods that have a polynomial running time if the data does not use "large" numbers. Such methods are called pseudo-polynomial. After that we will present an implementation of the first method, making it *strongly polynomial*, i.e. polynomial in the size of the network. Finally, we present two modified algorithms that use a

W. Hochstättler, A. Schliep, *CATBox*, DOI 10.1007/978-3-642-03822-8_7,
© Springer-Verlag Berlin Heidelberg 2010

scaling technique and yield efficient algorithms that have a polynomial time worst case behavior that also depends on the size of the numbers.

7.2 Optimality Criteria

We will develop our first optimality criterion by comparing two feasible solutions f and f'. Since $Bf = Bf' = b$ we necessarily must have $B(f - f') = 0$. The simplest non-zero integer solution that satisfies this flow conservation law at every vertex is a directed circuit.

In order to examine this situation in more detail we define the support graph of a flow.

Definition 20 Let $G = (V, A)$ be a directed graph and $g \in \mathbb{R}^A$ a (not necessarily non-negative) flow. The *support graph* of g is the graph

$$G(g) = (V, S), \text{ where } S := \{a \in A \mid g(a) > 0\} \cup \{-a \in -A \mid g(a) < 0\}.$$

Thus, if a is an arc in $G(f - f')$, where f and f' are feasible for (7.1), a is also an arc in $RN(D, f')$, that is $G(f - f')$ is a sub-digraph of $RN(D, f')$. If we associate the absolute value of $f - f'$ with each of these arcs we get a flow $\tilde{f} \geq 0$, that is *balanced* at each vertex, i.e. for all $v \in V$ we have flow conservation

$$\sum_{(u,v)\in A\cup-A} \tilde{f}(u, v) = \sum_{(v,u)\in A\cup-A} \tilde{f}(v, u).$$

Such a flow is also called a *circulation*. We will show that such a circulation is always a non-negative linear combination of incidence vectors of directed circuits.

Theorem 19 *Let f, f' be feasible flows for Problem 10 and $G(f - f') = (V, S)$ the support graph of $f - f'$. Then there exist directed circuits $C_1, \ldots, C_r, r \leq |S|$, in $G(f - f')$ and $\lambda_1, \ldots, \lambda_r \geq 0$ such that $\tilde{f} = |f - f'| = \sum_{i=1}^r \lambda_i \chi_{C_i}$.*

Proof We proceed by induction on $|S|$, the number of arcs in $G(f - f')$. If there are no arcs in that graph there is nothing to prove. Thus, let $e_1 = (u, v)$ be an arc with a positive flow. Since v is balanced there has to be an edge $e_2 = (v, w)$ leaving v with a positive flow. Similarly, we find an edge $e_3 = (w, \tilde{u})$. This way we can proceed, inductively. As the graph is finite we must have repeated vertices in that sequence and, thus, we find a directed circuit C_1 in $RN(D, f')$. Let $\lambda_1 = \min_{e \in C_1} \tilde{f}_e$. Then $\tilde{f}' := \tilde{f} - \lambda_1 \chi_{C_1}$ is a non-negative flow that is balanced at each vertex and the arc set of $G(\tilde{f}')$ is a proper subset of the arc set of $G(\tilde{f}) = G(f - f')$. By inductive assumption there exist $\lambda_2, \ldots, \lambda_r \geq 0$ where $r \leq |S|$ and directed circuits $C_2, \ldots, C_r$ such that $\tilde{f}' = \sum_{i=2}^r \lambda_i \chi_{C_i}$. $\qquad\square$

If a is an edge in a capacitated network and of cost $c_a \geq 0$, we associate a cost of $-c_a$ with the corresponding backward edge. This makes sense, since using this edge in a residual network corresponds to diminishing the flow sent through the forward edge. This way we define a cost function for the residual network $RN(D, f)$ of a problem and a flow. Now we can prove:

Theorem 20 (Circuit Optimality Criterion) *Let f be a feasible flow for Problem 10. Then f is optimal if and only if there exists no directed circuit C of negative cost $\sum_{a \in C} c_a$ in $RN(D, f)$.*

Proof First, assume C is a directed circuit in $RN(D, f)$ and $c(C) < 0$. In the same way in which we updated flow along an s-t-path in the maximal flow problem, we modify the flow along the circuit. Thus, let $\chi(C) \in \{0, \pm 1\}^A$ be the signed characteristic function of C, see (6.1). Let $f_1 = f + \min_{a \in C}\{rescap(a)\}\chi(C)$, where *rescap* denotes the residual capacity. For arbitrary $a \in A$ we have

$$f_1(a) \leq f(a) + rescap(a)\chi(C)(a) = f(a) + cap(a) - f(a) = cap(a)$$

and for $-a \in A$

$$f_1(a) \geq f(a) - f(-a) = 0.$$

Thus, f_1 is feasible and

$$c^\top f_1 = c^\top f + \min_{a \in C}\{rescap(a)\}c^\top \chi(C) < c^\top f.$$

On the other hand let f_1 be feasible for Problem 10 and $c^\top f_1 < c^\top f$. Write $f_1 - f$ as a conic combination of directed circuits in $G(f_1 - f)$. As $c^\top(f_1 - f) < 0$ is negative at least one of these circuits must correspond to a directed circuit of negative weight in $RN(D, f)$. $\square$

If f is optimal there is no negative circuit in $RN(D, f)$. As a consequence and by Sect. 5.5 we can define distances or potentials with respect to some arbitrary reference point $s \in V$. Let $S \subseteq V$ denote the set of vertices reachable on a directed path from s. For the fixed vertex s we set $pot(s) = 0$ and for all $v \in S$ we set $pot(v)$ to the length of a shortest, with respect to c, s-v-path. If this way vertices in $V \setminus S$ have an undefined potential because the directed cut $(S, V \setminus S) = \emptyset$. However, $(V \setminus S, S) = [S, V \setminus S] \neq \emptyset$ provided the underlying undirected graph is connected.

If we invert the direction of all arcs on this cut and set the cost of each such inverted arc $-a$ to $c_{-a} := -c_a$, the graph maintains the property of having no negative directed circuits, since no directed circuit may pass a directed cut. (Note, that a directed circuit that meets a cut in a forward edge must meet it in a backward edge as well). Proceeding this way we can define a distance label for all vertices in the network, yielding a potential function *pot* on $RN(D, f)$, that satisfies the inequality $pot(j) - pot(i) \leq c_{ij}$ for all arcs (i, j). Vice versa, such a potential function can exist only if there are no negative circuits, since summing up the inequalities along a circuit yields

$$0 \leq \sum_{e \in C} c_e.$$

Theorem 21 (Reduced Cost Optimality Criterion) *A feasible solution f for Problem 10 is optimal if and only if there exists a potential function pot for the vertices of the residual network $RN(D, f)$ satisfying*

$$c_{ij}^{pot} := c_{ij} + pot(i) - pot(j) \geq 0 \quad \text{for all } (i, j).$$

$\square$

We call $c_{ij}^{pot} := c_{ij} + pot(i) - pot(j)$ the *reduced cost*, as its value is the cost of re-routing one unit of flow from s to j via i. This is seen easiest by first considering $pot(i)$ as the cost of sending one unit of flow from s to i. Forwarding this unit from i to j we save $pot(j)$ and have to pay $c_{ij} + pot(i)$. Thus, the additional cost is $c_{ij} + pot(i) - pot(j)$.

This interpretation is valid not only for a single edge but for an arbitrary directed path.

Proposition 7 *Let pot be a potential function and c_{ij}^{pot} the reduced cost with respect to pot.*

(i) *If P is a directed path from v_k to v_l, then $\sum_{e \in P} c_e^{pot} = \sum_{e \in P} c_e + pot(v_k) - pot(v_l)$.*

(ii) *If C is a directed circuit then $\sum_{e \in C} c_e^{pot} = \sum_{e \in C} c_e$.*

Proof The inner potentials cancel. $\square$

Both of the above optimality criteria are formulated in terms of the residual network. The next theorem transfers the reduced cost criterion into the original network. From a linear programming point of view it is a special case of Theorem 5.

Theorem 22 (Complementarity) *A feasible solution f is an optimal solution of Problem 10 if and only if there exist node potentials pot such that its reduced costs satisfy:*

$$c_a^{pot} > 0 \Rightarrow f_a = 0$$
$$c_a^{pot} < 0 \Rightarrow f_a = cap(e)$$
$$0 < f_a < cap(a) \Rightarrow c_a^{pot} = 0.$$

Proof Assume f is an optimal solution and *pot* a node potential as given by Theorem 21. If $c_{ij}^{pot} > 0$, then by the Reduced Cost Criterion $RN(D, f)$ must not contain the backwards arc (j, i) for $c_{ji}^{pot} = -c_{ij}^{pot} < 0$. Thus $f_{ij} = 0$. Similarly, if $c_{ij}^{pot} < 0$, then $RN(D, f)$ cannot contain the arc (i, j). Hence it must be saturated meaning $f_{ij} = cap(i, j)$. In the last case the residual network contains both arcs implying $c_{ij}^{pot} \geq 0$ as well as $c_{ji}^{pot} \geq 0$ and hence $c_{ij}^{pot} = -c_{ji}^{pot} \leq 0$ which yields $c_{ij}^{pot} = 0$.

If on the other hand f is feasible and pot a node potential as in the theorem, then the residual network can contain both arcs a and $-a$ only if $c_a^{pot} = 0$. In the other two cases we must have $c_a^{pot} \geq 0$ and the claim follows.   □

In the forthcoming paragraph we will show that the above conditions are indeed the complementary slackness conditions of linear programming. More precisely, they are the conditions for a program with upper bounds on the variables, without explicit variables for the bound condition in the dual.

If we dualize the linear program for the minimum-cost flow problem we get

$$\max -y^\top cap + pot^\top b$$

such
that

$$-y^\top + pot^\top B \leq c^\top$$
$$y \geq 0.$$

The explicit conditions for complementary slackness as in Theorem 5 for our dual pair of linear programs are:

$$f_{ij} > 0 \Rightarrow -y_{ij} - pot(i) + pot(j) = c_{ij} \tag{7.2}$$
$$-y_{ij} - pot(i) + pot(j) < c_{ij} \Rightarrow f_{ij} = 0 \tag{7.3}$$
$$y_{ij} > 0 \Rightarrow f_{ij} = cap_{ij} \tag{7.4}$$
$$f_{ij} < cap_{ij} \Rightarrow y_{ij} = 0 \tag{7.5}$$

If we compare this with the conditions for complementarity in the last paragraph we see that here in addition we have edge variables y. A closer examination will reveal that these are not necessary for our optimality criterion.

First note, that the constraint $-y^\top + pot^\top B \leq c^\top$ of the dual program might as well be written as $y \geq -c^{pot}$. Since there are no further constraints on y and its coefficients in the program and the capacities cap are non-negative, y_{ij} will take the smallest possible value in an optimal solution, thus

$$y_a = \max\{0, -c_a^{pot}\}.$$

Second, if we combine condition (7.2) and (7.4) we see that $f_{ij} > 0$ implies, either $y_{ij} = 0$, which corresponds to the last case in Theorem 22, or $f_{ij} = cap_{ij}$. Thus, we may rewrite the conditions to become

$$cap_{ij} > f_{ij} > 0 \Rightarrow y_{ij} = c_{ij}^{pot} = 0$$
$$y_{ij} > -c_{ij}^{pot} \Leftrightarrow c_{ij}^{pot} > 0 \Rightarrow f_{ij} = 0$$
$$c_{ij}^{pot} < 0 \Rightarrow f_{ij} = cap_{ij}$$
$$f_{ij} < cap_{ij} \Rightarrow c_{ij}^{pot} \geq 0.$$

The last condition is implied by the others, thus we have derived the conditions of our theorem. Vice versa we can derive the conditions for complementary slackness from the conditions in Theorem 22 by setting $y_e = \max\{0, -c_e^{pot}\}$.

Before presenting our first algorithm to compute a minimum-cost flow, we want to point out that, similarly to the possibility to compute node potentials from an optimal flow using a shortest path algorithm, we can also compute a minimum-cost flow given an optimal node potential, if the data does not have a directed circuit of negative cost. Actually, without loss of generality it is possible to assume that all costs are integer, doing some transformations of the network. We will leave the details as an exercise. Interestingly enough, given the node potentials, the computation of a minimum cost flow is done using a maximum flow algorithm, namely as follows:

Let c_{ij}^{pot} denote the reduced costs of an optimal node potential. We use these to partition the edges in the three natural classes

$c_{ij}^{pot} > 0$: No optimal flow may use this edge, delete it.

$c_{ij}^{pot} = 0$: Leave the edge as it is.

$c_{ij}^{pot} < 0$: Any optimal flow has to saturate this edge. Thus, delete the edge, increase $b(j)$ by cap_{ij} and decrease $b(i)$ by the same amount.

This way we get a modified network with a supply and demand function b' where $\sum_{v \in V} b'(v) = 0$. Connecting the supply vertices v $(b(v) > 0)$ with a newly introduced source s by edges of capacity $b(v)$ and the demand vertices analogously with a new sink t, we construct a capacitated network where we solve a maximum flow problem (cf. Exercise 47). By construction, if the minimum-cost flow problem has a feasible solution, any optimal solution of the maximum flow problem is easily modified to a solution of the minimum-cost problem by adding the saturated edges. This solution satisfies the reduced cost criterion and, thus, is optimal.

Exercise 51 (i) We consider a flow problem with lower bound restrictions on the flow, i.e. where the restriction $0 \le f \le cap$ is replaced by $low \le f \le upp$.

$$\min c^\top f$$
$$Bf = b$$
$$low \le f \le upp$$

Transform this minimum cost flow problem into a standard one, i.e. with non-negativity constraints as lower bounds.

(ii) Let $G = (V, A, cap, c)$ be a standard instance of a minimum cost-flow problem. Transform a problem where some of the cost coefficients are negative, into a mincost flow problem with lower and upper bounds and non-negative cost.

7.3 First Algorithms

7.3.1 Canceling Negative Circuits, the Primal Method

We derive our first algorithm from the circuit optimality criterion. The algorithm follows the proof of Theorem 20. There we had seen how a negative circuit in the residual network can be used to decrease the cost of a flow. Thus our algorithm first computes a feasible flow by solving a maximal flow problem analogously to the method at the end of the last paragraph. Then, we continue improving the solution as long as we find negative circuits in the residual network.

This yields the following algorithm.

ALGORITHM NegativeCircuit

```
if EstablishFeasibleFlow():
    ready = False
    while not ready:
        C = FindNegativeCycle()
        if C == None:
            ready = True
        else:
            delta = MinimalRestCapacity(C)
            for (u,v) in Edges(C):
                if ForwardEdge(u,v):
                    flow[(u,v)] = flow[(u,v)] + delta
                else:
                    flow[(v,u)] = flow[(v,u)] - delta
    ShowTree(R,RA)
```

First, we compute a feasible flow using the reduction from Exercise 47, then we search for negative circuits to improve the solution. The algorithm finishes, when there is no negative circuit in the residual network. As a certificate for optimality we compute a shortest path tree. This is automatically done by a label correcting algorithm for a shortest path problem which we use in order to detect a negative circuit as discussed in Sect. 5.6.

Theorem 23 *If there exists a feasible flow for the problem, the algorithm above finds a minimum-cost flow in $O(|V| \cdot |A|^2 CW)$, where C and W are the largest numbers in absolute values that occur in the cost function respectively the capacity function. If all capacities are integer, so is an optimal solution.*

Proof   Setting $y_a = \max\{0, -c_a\}$ we find a feasible solution for the dual program of value $\geq -|A|CW$. This is a lower bound for the primal problem. A feasible flow the cost of which is bounded by $|A|CW$ is found in $O(|V||A|^2)$ by solving a maximum flow problem. In each iteration the cost of the flow decreases by at least one unit. Therefore the algorithm goes through at most $2|A|CW$ iterations. A negative circuit is detected at a cost of $O(|V||A|)$. Update of the data is done in linear time. Clearly all integer data will remain integer throughout the algorithm.   □

Remark 10 In general, the algorithm above has no running time bound polynomial in the size of the input, similar to the algorithm of Ford and Fulkerson, since the numbers and not their coding length appear in the runtime. Such a behavior is called pseudo-polynomial. By a clever choice of the negative circuits a polynomial running time behavior can be achieved. We will discuss such an algorithm, that can be considered a generalization of Edmonds-Karp in Sect. 7.5.

7.3.2 Augmenting Shortest Paths, the Dual Method

The Circuit Cancellation Algorithm is a primal algorithm in the sense that it always keeps a feasible flow that is improved in every step. In this paragraph we follow the dual paradigm and always keep dually feasible node potentials. By Exercise 51 we may assume that the cost function is non-negative and thus, an initial node potential can be easily computed. We always keep a pseudoflow, which is complementary to the potential, and try to turn that into a flow. The precise definition of a pseudoflow is:

Definition 21 Let $D = (V, A, cap)$, $c : A \to \mathbb{Z}$ and $b : V \to \mathbb{Z}$ be an instance of Problem 10. A function $f : A \to \mathbb{Z}_+$ satisfying $f \le cap$ is called a *pseudo flow*. The *imbalance* $ex(v)$ of a vertex $v \in V$ is defined as

$$ex(v) = b(v) + \sum_{(u,v) \in A} f(u, v) - \sum_{(v,u) \in A} f(v, u).$$

Thus the imbalance is the signed deviation of the current flow into vertex v from is demand or supply.

By adjusting the supply and demand vector we can consider any pseudoflow an optimal flow for a properly defined problem. Moreover, using the distances computed with respect to non-negative reduced costs, which are defined with respect to a given potential we can send flow along a shortest path and, using the distances from a given vertex s to all other vertices we can update the potential such that the reduced costs remain non-negative. But such a vertex need not exist. Since flow will only be sent from demand vertices, only vertices reachable from some demand vertex are relevant for our task. Hence, we can add a super source s from which all supply vertices are reachable at no cost with a capacity equal to their supply. Using the distances from this vertex s we can prove:

Lemma 17 Let f be a pseudo flow and pot a node potential, such that f and pot are a dual pair of optimal solutions for the problem where the vector b has been replaced by $b' = b - ex(D, b)$.

Let $dist$ denote the distances from some fixed vertex s to all other vertices with respect to to the reduced cost c_{ij}^{pot}. Then the following holds:

(i) $pot' := pot + dist$ is a dually optimal node potential as well.
(ii) The reduced cost $c_{ij}^{pot'}$ is zero on the edges of the shortest-path-tree starting in s.

Proof By assumption $c_{ij}^{pot} \geq 0$ for all edges in $RN(D, f)$ (with respect to the modified problem). Since *dist* is a distance vector, it respects the condition

$$dist(j) - dist(i) \leq c_{ij}^{pot}.$$

This implies

$$
\begin{aligned}
c_{ij}^{pot'} &= c_{ij} + pot'(i) - pot'(j) \\
&= c_{ij} + pot(i) + dist(i) - pot(j) - dist(j) \\
&= c_{ij} + pot(i) - pot(j) + (dist(i) - dist(j)) \\
&\geq c_{ij}^{pot} - c_{ij}^{pot} = 0.
\end{aligned}
$$

Hence *pot'* still is a dually optimal node potential for the problem defined by b'. Note that the last inequality also holds if i is not reachable from s and hence $dist(i) = \infty$.

For a proof of the second assertion we consider an edge on a shortest path. There, necessarily, we have $dist(j) - dist(i) = c_{ij}^{pot}$, thus the last inequality is in fact an equality. $\qquad\square$

As a consequence of these considerations we get

Corollary 7 *Let f be a pseudo flow satisfying the reduced cost criterion. Assume that f' arises from f by an augmentation along some shortest s-t path. Then f' satisfies the reduced cost criterion as well.*

Proof Replacing *pot* by *pot'* as defined in the last lemma, the reduced cost along any shortest path is zero. Therefore, the augmented pseudo flow satisfies the reduced cost criterion as the flow was not changed anywhere else. $\qquad\square$

The last lemma suggests the following algorithm. As a preparatory step, in order to detect infeasible instances, we introduce an additional edge of infinite capacity and cost for any two vertices. Clearly, such an edge cannot occur in any optimal solution of a feasible instance.

Initially, we set the node potential to constantly zero. This is complementary to the zero flow. Next, we compute shortest path distances from our artificial super source and choose a sink t, with negative imbalance, and send a maximal amount of flow along an s-t-path P that is shortest with respect to the reduced cost, taking into account the imbalance of the supply vertex s' adjacent to s and the imbalance of t as well as the capacity restrictions along P. We complete the iteration by adding the distance function to the potential.

ALGORITHM SuccessiveShortestPath

```
5  while D.IsNotEmpty():
       t = D.Pop()
       ShortestPathDist()
       (s,P) = FindShortestPath(t)
       delta = min(excess(s),-excess(t),MinResCap(P))
10     IncreaseFlow(P,delta)
       LiftVertices(dist)
       if excess(t) < 0:
           D.Push(t)
```

According to our assumptions on the artificial edges there will always be a path from s to t in the residual network. In the beginning the distance function $pot = dist$ derived from the edge costs is a node potential that satisfies the reduced cost criterion with respect to the zero flow. By Corollary 7 this property is maintained throughout the algorithm while the demand decreases. Thus, in finite time we construct a feasible flow that satisfies the reduced cost criterion.

The number of iterations is bounded by $B := \frac{1}{2}\sum_{v \in V}|b(v)|$. The most costly computation in each iteration is the solution of the shortest path problem. Thus, the overall complexity is $O(B\,S(|A|, |V|, |V|C))$, where $S(|A|, |V|, |V|C)$ is the complexity of a shortest path algorithm and the total edge costs are bounded by $|V|C$. Here C is an upper bound on the absolute value of the cost of a single edge. This bound can also be exploited to find the proper value of infinity for the artificial edges.

7.4 The Primal Dual Method

As we have seen that the minimum-cost flow problem reduces to a maximal flow problem in the network of the edges of reduced cost zero with respect to an optimal node potential, we might as well consider the following primal dual approach.

ALGORITHM Iterate until a feasible flow has been found:

- Compute the distance labels with respect to reduced costs and update $pot = pot + dist$.
- Solve a maximum flow problem with multiple sources and sinks on the graph of the tight edges, i.e. the edges with reduced cost zero.

Here we will verify only the correctness of this approach. Recall our dual pair of linear programs.

$$
\begin{array}{ll}
\min c^\top f & \\
\text{subject} \quad Bf = b & \\
\text{to} \qquad\quad f \le cap & \\
\qquad\qquad f \ge \ 0. &
\end{array}
\qquad
\begin{array}{ll}
\max -y^\top cap + pot^\top b & \\
\text{subject} \quad -y + pot^\top B \le c & \\
\text{to} \qquad\qquad\qquad y \ge 0. &
\end{array}
$$

Basically the above approach does several augmentations along shortest paths in one go. When we proved that the reduced cost criterion remains valid after an update on a shortest path we actually showed, that any modification along edges of zero reduced cost preserves this property. Thus the correctness of this approach follows.

Software Exercise 52 Load the algorithm `NegativeCircuit.alg` and the graph `MCF4to1A.cat`. First we compute a feasible flow by adding an artificial source and sink, that are connected to the sources and the sinks by edges with a capacity that equals their supply respectively demand. A maximal flow that saturates all the new edges yields a feasible flow. If such a flow does not exist, we will find a cut proving infeasibility of the instance.

Now our feasible flow basically sends 75 units along the upper boundary of our graph and 25 right through the middle. The first negative circuit of cost -9 that is found reroutes flow from 8-6-13 to 8-28-30-13. The second also modifies only the "flow through the middle". The third and the fourth negative circuits are quite large and reroute 25 units each from the upper to the lower boundary. Finally we reroute 16 units from the upper boundary through the middle and terminate after 5 iterations giving shortest path tree (green edges) and the distances from vertex 12 to all vertices as certificate that there are no more negative circuits in the residual network.

Now we try the dual approach on the same instance and load the algorithm SuccessiveShortestPath.alg. In the beginning all vertices have a potential of zero. The path of least cost from our invisible super source s—you can try to detect it in the procedure ShortestPathDist() in SuccessiveShortest Path.pro—to 12 sends supply from 20 along the bottom of the graph. We send 25 units of flow and complete the iteration by updating the potential and the reduced costs in LiftVertices(dist). In the residual graph we can now distinguish edges of positive (red), zero (grey) or negative (green) cost by their color. Recall that we may have a non-zero flow only on edges with non-negative reduced costs, and that an arc with a positive reduced cost must be saturated. Fortunately the picture is compatible with our theory and there are no (green) edges of negative reduced cost. Using the shortest path from s to 12 with the updated distances we send 25 units of flow from vertex 2. In the next iteration the shortest path has a bottleneck at arc (30,13) which has a residual capacity of only 24. In iteration 4 we find a path from vertex 1, but the residual capacity of (13,12) allows for only a flow of 9. Now the remaining supply from 23 is sent along the top of the network and, finally, a long and winding path in the residual network enables as to redirect some flow originating in 2 and 23 and to compute a feasible and hence optimal flow.

After six iterations in total we end up with the same solution as computed by the Circuit Canceling algorithm.

7.5 Polynomial Time Algorithms

All of the algorithms of the last section have only a pseudo-polynomial running time bound, meaning that a runtime polynomial in the size of the input data can no longer be guaranteed if the capacities or the supply or demand become big (say $O(2^{|A|})$).

7.5.1 Min Mean Circuit Canceling

Motivated by successful approaches to design *strongly polynomial* primal algorithms for the maximum flow problem, that is algorithms where the number of

iterations is bounded by a polynomial in the size of the network, we want to look
for "good" negative circuits for the Circuit Canceling algorithm. Probably there is
no polynomial time strategy to find a minimal circuit of maximal absolute value,
i.e. a circuit that yields the largest decrease in our objective function, as this would
require a polynomial time strategy to solve among others the Hamiltonian Circuit
problem, which is well-known to be $\mathcal{NP}$-complete.

One strategy analogous to the Edmonds-Karp implementation of the algorithm of
Ford and Fulkerson would require to update on a negative circuit with as few edges
as possible. As far as we know, this has not been successfully pursued. Goldberg
and Tarjan proved in 1989 that a more direct analogy, a minimum mean circuit
augmentation, yields a strongly polynomial time algorithm.

In each iteration we choose a negative directed circuit in $RN(D, f)$ that minimizes

$$-\gamma := \sum_{e \in C} \frac{c_e}{|C|}. \tag{7.6}$$

We can model the maximum flow problem by introducing a backward arc of infinite capacity and cost -1, whereas all other arcs have zero cost. The corresponding
min-cost-flow problem with $b = 0$ is equivalent to the maximum flow problem. The
min-mean-circuit implementation in this case coincides with Edmonds-Karp.

In the following we will analyze the dynamics of the parameter γ_i in order to
show that the number of circuit cancellations is bounded by a polynomial in the size
of the network.

Let $C_1, \ldots, C_k$ be the augmenting circuits and $\gamma_1, \ldots, \gamma_k$ the corresponding
absolute values of the mean costs. Then γ_i is the smallest positive number with
the property that adding γ_i to the cost of each arc in the residual network makes all
negative circuits disappear or equivalently such that there exists a potential function
pot^i satisfying $c_a + \gamma_i \geq pot^i(v) - pot^i(u)$ for all arcs $a = (u, v)$ in $RN(D, f)$.

Lemma 18 *If the parameters $\gamma_1, \ldots, \gamma_k$ are defined as above and using $m = |A|$
and $n = |V|$ we have:*

(i) $\gamma_{i+1} \leq \gamma_i$
(ii) $\gamma_{i+m} \leq \frac{n-1}{n} \gamma_i.$

Proof By definition and the above remark there exists a potential function pot^i
such that for all arcs $a = (u, v) \in RN(D, f)$ we have

$$c_a^{pot^i} = c_a + pot^i(u) - pot^i(v) \geq -\gamma_i. \tag{7.7}$$

By definition of γ_i this necessarily implies that we have equality on all arcs of C_i.
The only new arcs that are possibly introduced in $RN(D, f)$ after the update on
C_i are the backward arcs for C_i with reduced cost $\gamma_i > 0$. Thus (7.7) still holds in

the updated residual network. As γ_{i+1} is the smallest positive number such that a potential exists satisfying $c^{pot^{i+1}}(a) \geq -\gamma_{i+1}$ this implies the first part.

For the second assertion we consider the sequence $C_i, \ldots, C_{i+m-1}$ of augmenting minimum mean circuits. We claim that one of theses circuits must contain an arc of non-negative cost. Assume this were not the case. Then each of the augmentation steps saturates an arc of negative cost and the corresponding backward arc is not used by any of the following circuits. As there are only $|A| = m$ arcs, this is impossible. So let h be the smallest non-negative integer such that C_{i+h} contains an arc a_0 such that $c_{a_0} \geq 0$. Arcs of negative cost are newly introduced in $RN(D, f)$ only if the flow is augmented along an arc of positive cost. Hence all negative arcs a in C_{i+h} were already arcs in the residual network before the i-th augmentation and satisfy $c_a^{pot^i} \geq -\gamma_i$. Thus

$$\sum_{a \in C_{i+h}} c_a = \sum_{a \in C_{i+h}} c_a^{pot^i} \geq -(|C_{i+h}| - 1)\gamma_i$$

$$\Rightarrow \gamma_{i+h} = - \sum_{a \in C_{i+h}} \frac{c_a}{|C_{i+h}|} \leq \frac{|C_{h+i}| - 1}{|C_{h+i}|} \gamma_i \leq \frac{n-1}{n} \gamma_i$$

and the assertion follows from the first part of the lemma.   □

Now, let $t = 2mn \lceil \ln(n) \rceil$. We will show that at least every t iterations we can fix an arc to its current flow value for the remainder of the algorithm:

Lemma 19 *For all i such that the algorithm does not terminate within the next t steps there is an arc $a \in C_i$ such that the flow on a does not change in any step h where $h \geq i + t$.*

Proof   The following computation verifies that $\gamma_{i+t} < \frac{\gamma_i}{2n}$.

$$\frac{\gamma_{i+t}}{\gamma_i} \leq \left(\frac{n-1}{n}\right)^{2n \lceil \ln(n) \rceil}$$

$$\leq \left(\left(1 - \frac{1}{n}\right)^n\right)^{2 \ln(n)}$$

$$< \left(\frac{1}{e}\right)^{2 \ln(n)} = \frac{1}{n^2} \leq \frac{1}{2n}.$$

Now let $a_0 \in C_i$ such that $c_{a_0}^{pot^{i+t}} \leq -\gamma_i < -2n\gamma_{i+t}$. Such an a_0 exists as $\sum_{a \in C_i} c_a = \sum_{a \in C_i} c_a^{pot^{i+t}}$. Now $c_{a_0}^{pot^{i+t}} < -2n\gamma_{i+t} \leq -\gamma_{i+t}$ thus a_0 cannot be an arc in the residual network in step $t + i$. So it must be saturated. Assume there exists a step $h > t + i$ such that the flow on a_0 is decreased again.

Thus, if f_h respectively f_{i+t} are the flows in the the h-th respectively the $i + t$-th step then applying Theorem 19 (see page 90) to $f_{i+t} - f_h$ we find a directed circuit

C_0 using a_0 in $RN(D, f_h)$ such that its reversion $-C_0$ is a directed circuit in $RN(D, f_{i+t})$. Hence $\forall a \in C_0 : -c_a^{pot^{i+t}} = c_{-a}^{pot^{i+t}} \geq -\gamma_{i+t}$ and thus

$$\sum_{a \in C_0} c_a = \sum_{a \in C_0} c_a^{pot^{i+t}} = c_{a_0}^{pot^{i+t}} + \sum_{a \in C_0 \setminus \{a_0\}} c_a^{pot^{i+t}}$$

$$< -2n\gamma_{i+t} + (|C_0| - 1)\gamma_{i+t} \leq -n\gamma_{i+t} \leq -|C|\gamma_h,$$

which contradicts the definition of γ_h.   □

Thus every t iterations at least one new edge t is fixed and the algorithm has to terminate after $mt = O(m^2 n \ln(n))$ iterations.

Theorem 24 *The Mean Min Circuit Canceling algorithm terminates after $O(m^2 n \ln(n))$ iterations.*   □

Thus, we have a strongly polynomial time implementation once we can find a minimum mean circuit in polynomial time. We will do this by dynamic programming. Recall that dynamic programming for shortest paths failed in the presence of negative circuits. It did not help to add a constant to all costs in order to make all negative circuits disappear, since this might change the shortest path, i.e. a path using less edges might become shorter this way. This is no longer the case when we consider minimum mean circuits, the values of which will all change by the same constant. The basic result for our algorithm, which is due to Karp, is the following result.

Theorem 25 *Let $s \in V$ such that for all $v \in V$ there exists a directed s-v-path. Let γ be as in (7.6). For all $v \in V$, $1 \leq k \leq n$ let $d_k(v)$ denote the length of a shortest $s - v$ walk using exactly k edges with respect to c, and $d_k(v) = \infty$ if no such path exists. Then*

$$-\gamma = \min_{v \in V} \max_{1 \leq k \leq n-1} \frac{d_n(v) - d_k(v)}{n - k}. \tag{7.8}$$

Proof We consider the problem with respect to the modified cost function $\tilde{c} := c + \gamma$. The corresponding distance function $\tilde{d}$ satisfies

$$\tilde{d}_k(v) = d_k(v) + k\gamma. \tag{7.9}$$

By the definition of γ, the network with respect to the cost function $\tilde{c}$ does not have any negative cycle. Thus, for any $v \in V$ there is an acyclic shortest $s - v$ path and hence some $1 \leq k \leq n - 1$ such that $\tilde{d}_n(v) \geq \tilde{d}_k(v)$ implying

$$\frac{d_n(v) - d_s(v)}{n - k} = \frac{\tilde{d}_n(v) - n\gamma - \tilde{d}_k(v) + k\gamma}{n - k} \geq -\gamma.$$

To prove the other inequality let C be a circuit where $-\gamma$ is attained in (7.6). Let $w \in C$ and $\tilde{P}$ a shortest s-w-path with respect to $\tilde{c}$. Assume that w is chosen such that $\widetilde{P}$ uses as few edges as possible. Then $|\widetilde{P}| \leq n - |C|$. We extend $\widetilde{P}$ to a walk P of n edges traversing the edges of C. Let v be the endpoint of P. Note, that as there is no negative circuit with respect to $\tilde{c}$, C has weight 0 and $\widetilde{P}$ is a shortest s-w-path, P necessarily is a shortest walk from s to v. Thus, $\tilde{d}_n(v) \leq \tilde{d}_k(v)$ and

$$\frac{d_n(v) - d_k(v)}{n - k} = \frac{\tilde{d}_n(v) - n\gamma - \tilde{d}_k(v) + k\gamma}{n - k} \leq -\gamma.$$

$\square$

It is straightforward to compute γ and a minimum mean circuit using dynamic programming. For that purpose we use the recursion

$$d_{k+1}(v) = \min\{d_k(u) + c((u, v)) \mid (u, v) \in A\} \tag{7.10}$$

and modify the Bellman-Ford algorithm accordingly. The circuit is found using a pointer structure similar to the algorithm described in Chap. 5.

Exercise 53 Write a routine that finds a circuit of minimum mean weight in a weighted digraph.

Software Exercise 54 We included an implementation of such a routine in the prologue of the Circuit Canceling algorithm. The routine is called `MinMeanCycle()`. If you replace line 7: `C=FindNegativeCycle()` with `C=MinMeanCycle()` you can explore the similarities and differences between the pseudo-polynomial and the strongly polynomial time implementation. Most probably your observation will be that the asymptotically inferior behavior of the former is not visible in instances that can be treated with our software.

In practice there is another important concept, namely scaling, which is usually applied to the successive shortest path algorithms.

Although there exists a scaling algorithm by Orlin which yields a strongly polynomial running time, here we will introduce more simple versions that are "only" polynomial time. So, the concept is to scale and round the data and do a logarithmic number of computations of rounded problems to iteratively reach the exact optimum. We will consider two of these algorithms. In both cases the logarithms of the numbers are still a parameter in the running time bound, so the algorithms are not strongly polynomial. Recall that all polynomial time algorithms in the former chapters had such a behavior.

7.5.2 Capacity Rounding

The idea to turn the successive shortest path algorithm into a polynomial one actually is quite simple. We want to avoid augmentations that are relatively small compared to the size of the data. For that purpose we introduce a lower bound Δ on the size of the augmentations. If we simply delete all edges of residual capacity $< \Delta$ we can guarantee this lower bound for any augmentation and can proceed by doing shortest path augmentations in that network. We iterate this process of deleting edges of small capacity and augmenting until no other augmentation $\geq \Delta$ is found. Since our artificial edges have infinite capacity this can only mean that there either no longer exists a supply node supplying or no demand node demanding at least Δ units. In particular the sum B of supply or demand is bounded by $|V|\Delta$.

Of course, the reduced cost criterion may be violated for the deleted edges and thus we may have edges where $c_{ij}^{pot} < 0$. We fix this temporarily by saturating these edges and modifying supply and demand at their end vertices accordingly. The parameter B by this procedure may grow up to $(|V| + |A|)\Delta$. Our node potential now satisfies the reduced cost criterion, we may divide Δ by 2 and iterate. All in all we get the following procedure

– Let $\Delta = 2^{\lfloor \log(C) \rfloor}$.
– While $\Delta \geq 1$
– Set $\tilde{w} = \lfloor \frac{w}{\Delta} \rfloor$, $\tilde{b} = \lfloor \frac{b}{\Delta} \rfloor - r$. Solve the modified minimum-cost flow problem starting from flow f using successive shortest paths.
– Fix edges of small residual capacity $\leq \Delta$ yielding a correction vector r and augment f.
– $\Delta = \frac{\Delta}{2}$.
– Iterate.

Analyzing the runtime of this procedure we see that we do $\log C$ iterations. We may assume that $\log B$ and $\log C$ in the beginning have the same order of magnitude (for a feasible flow to exist). At the end of the first iteration the size of B is bounded by $(|A| + |V|)\Delta$. Thus, we have a running time bound of $O((|A| + |V|)S(|A|, |V|, |V|C))$ per iteration and

$$O((|A| + |V|) \log(C) S(|A|, |V|, |V|C))$$

for the full algorithm.

We did not include an implementation in our software. We did not find small examples that demonstrate the superiority of these polynomial time algorithms to the pseudo polynomial time algorithms. The same holds for the implementation of the algorithm in the next paragraph. You can convince yourself and search for such an example.

7.5.3 Cost Scaling

While we ignored the lower bits of the capacity function in the previous section, here we will operate on the bits of the cost function. Again we will exploit the fact that we know the optimal potentials for the rounded problem. This turns the minimum-cost flow problem into a maximum flow problem which we can solve with the preflow push algorithm. But this is just the basic idea and it needs to be implemented carefully. We start with some definitions. The first one defines an ε-optimal flow as a flow that approximately satisfies the conditions from Theorem 22.

Definition 22 Let $\varepsilon > 0$. A pseudo flow is called *ε-optimal*, if there exist node potentials pot such that the following ε-approximations of the complementarity conditions are satisfied:

$$c_a^{pot} > \varepsilon \Rightarrow f_a = 0$$
$$c_a^{pot} < -\varepsilon \Rightarrow f_a = cap_a$$
$$0 < f_a < cap_a \Rightarrow |c_a^{pot}| \leq \varepsilon.$$

Summarizing we can write the conditions as

$c_a^{pot} \geq -\varepsilon$ in RN(D,f).

Lemma 20 *If $\varepsilon > C$, then any feasible flow is ε-optimal. If on the other hand $\varepsilon < \frac{1}{|V|}$, then any ε-optimal flow is optimal.*

Proof If f is feasible, put $pot = 0$. Since C is an upper bound on the absolute value of the cost of all edges $|c_a^{pot}| \leq C$ and the conditions are satisfied trivially. If on the other hand $\varepsilon < \frac{1}{|V|}$, then let W be a directed circuit in $RN(D, f)$. As any edge in the residual network satisfies $|c_a^{pot}| \leq \varepsilon$, we get $c(W) = c^{pot}(W) \geq -|V|\varepsilon > -1$. Thus, as all data is integer the residual network can no longer contain a negative directed circuit. $\square$

In our algorithm we always keep an $\frac{\varepsilon}{2}$-optimal pseudoflow which we turn into a flow using the node potentials to control a preflow-push type operation. From this flow we construct an $\frac{\varepsilon}{4}$-optimal pseudo flow, divide ε by 2 and iterate until ε is small enough. The flow is modified only along *feasible edges*, i.e. edges of reduced cost satisfying

$$-\frac{\varepsilon}{2} \leq c_{ij}^{pot} < 0.$$

In order to keep ε integer throughout the algorithm we multiply all data with $|V| + 1$ and initially set

$$\varepsilon := 2^{\lceil \ln(|V|C+1) \rceil}.$$

This yields the following algorithm:

ALGORITHM CostScaling

```
     eps = RecalculateCosts() # for integer eps
11
     while eps > 1:
         for e in Edges():
             if ReducedCosts(e) > 0:
                 flow[e] = 0
16           elif ReducedCosts(e) < 0:
                 flow[e] = cap(e)

         IsFlow = False
         while not IsFlow:
21               v = ActiveVertex()
             if v:
                 e = AdmissibleEdge(v,eps)
                 if e:
                     delta = Min(res(e),excess(v))
26                   if ForwardEdge(e[0],e[1]):
                         flow[(e[0],e[1])] = flow[(e[0],e[1])] + delta
                     else:
                         flow[(e[1],e[0])] = flow[(e[1],e[0])] - delta
                 else:
31                   Lower(v,eps/2)
             else:
                 IsFlow = True
         eps = eps/2
```

For a proof of correctness and a running time analysis of this algorithm we show:

Lemma 21 *During execution of the inner while loop* (f, pot) *always forms an $\frac{\varepsilon}{2}$-optimal pseudo flow and an $\frac{\varepsilon}{2}$-optimal flow whenever IsFlow is set True.*

Proof We proceed by induction on the steps of the algorithm. Before the inner while loop is entered we construct a pseudo flow and enforce the optimality conditions to hold. Since the pseudo flow is even optimal, in particular it is $\frac{\varepsilon}{2}$-optimal. We perform a push on edge (i, j) only if it is feasible (admissible), i.e. if $-\frac{\varepsilon}{2} \leq c_{ij}^{pot} < 0$ and thus $c_{ji}^{pot} > 0$. Thus, any new backward arc introduced into the residual network in this way satisfies the criterion.

If we relabel an active vertex there was no feasible edge incident with this vertex anymore, i.e. no edge with non-zero residual capacity satisfies $-\frac{\varepsilon}{2} \leq c_{ij}^{pot} < 0$. Using this and the second inequality of $\frac{\varepsilon}{2}$-optimality of the pseudo flow we find that $c_{ij}^{pot} \geq 0$ for all edges with non-zero residual capacity leaving i. If we decrease the potential of i by $\frac{\varepsilon}{2}$, the reduced costs still will satisfy $c_{ij}^{pot} \geq \frac{\varepsilon}{2}$ preserving $\frac{\varepsilon}{2}$-optimality. $\square$

The last lemma clearly implies correctness of the algorithm. For the running time analysis we note:

Lemma 22 *During one execution of the inner while loop the node potential of each vertex is changed at most $3|V|$-times.*

Proof To see this, consider an $\frac{\varepsilon}{2}$-optimal pseudo flow f with respect to a potential pot. Let f' be the ε-optimal flow that had been computed when the inner while loop was exited in the former execution of the inner while loop and let pot' denote the corresponding potential. Since f' is a flow, for each vertex v with positive excess $ex(v) > 0$ in $RN(D, f)$ there exists some $w \in V$ such that $ex(w) < 0$. Moreover, we have a directed path $P = (v = v_0, v_1 \ldots v_k = w)$ in $RN(D, f)$, such that $v_k, v_{k-1} \ldots v_0$ is a directed path in $RN(D, f')$. By $\frac{\varepsilon}{2}$-optimality we conclude

$$\sum_{(i,j)\in P} c_{ij}^{pot} \geq -k\frac{\varepsilon}{2}$$

implying

$$\sum_{(i,j)\in P} c_{ij} \geq -k\frac{\varepsilon}{2} - pot(v) + pot(w). \tag{7.11}$$

Similarly, ε-optimality of pot' implies

$$\sum_{(i,j)\in P} c_{ji} \geq -k\varepsilon - pot'(w) + pot'(v). \tag{7.12}$$

As w is a demand vertex, its label has not been changed during the current pass through the while-loop, thus $pot(w) = pot'(w)$. Furthermore, using $c_{ij} = -c_{ji}$ addition of (7.11) and (7.12) yields

$$pot'(v) - pot(v) \leq k\frac{3\varepsilon}{2}, \tag{7.13}$$

implying the assertion.   □

Lemma 23 *In each pass through the inner while loop we have* $O(|V||A|)$ *saturating pushes.*

Proof We will show that another saturating push along an edge can only occur if both node potentials $pot(i)$ and $pot(j)$ have been decreased by at least $\frac{\varepsilon}{2}$. Consider the first saturating push along edge (i, j). This edge is feasible and thus $c_{ij}^{pot} < 0$. Before it is possible to saturate this edge again it has to be used in the opposite direction. This is only possible if at that point $c_{ij}^{pot} > 0$. Since node potentials decrease monotonically we conclude that $pot(j)$ must have decreased. For another use of (i, j) we have also to modify $pot(i)$. Thus, by last lemma we have at most $O(|V|)$ saturating pushes per arc which implies the assertion.   □

As in the case of the preflow push the hardest part of the analysis is the bound for the non-saturating pushes. Let us denote by *feasible network* the network consisting only of feasible edges.

Lemma 24 *The feasible network is acyclic while passing through the inner while loop.*

Proof We proceed by induction on the algorithm. In the beginning of the while loop the pseudo flow is 0-optimal and the residual network does not contain any feasible edge. A push on (i, j) implies that $c_{ij}^{pot} < 0$. Thus, if we introduce a new backward arc we have $c_{ji}^{pot} > 0$, implying that this edge will not be feasible and no directed circuit can appear. A relabel operation decreases the reduced cost of all arcs leaving i by $\varepsilon/2$ and may give rise to new feasible edges. On the other hand the reduced cost of all incoming arcs is increased by $\varepsilon/2$ making them infeasible. Thus, after a relabel operation of vertex i no directed circuit will pass i.   □

This enables us to prove

Lemma 25 *In each pass of the inner while loop the number of non-saturating pushes is $O(|V|^2|A|)$.*

Again we use a potential function. For each vertex i let $g(i)$ denote the number of vertices reachable from i on a directed path in the network of feasible edges and

$$\Phi := \sum_{i \text{ is active}} g(i)$$

the potential function. In the beginning, as there are no feasible edges at all, we have $\Phi \le |V|$. A saturating push along (i, j) may turn j into a new active vertex, and thus increase the potential function by at most $|V|$. By Lemma 23 the total increase of Φ caused by saturating pushes is $O(|V^2||A|)$. A relabel operation may cause new feasible edges and increase $g(i)$ by at most $|V|$. Since all incoming arcs incident with vertex i become infeasible all other values $g(j)$ can only decrease. Hence by Lemma 22 the total increase of the potential function caused by relabel operations is $O(|V|^3)$. A non-saturating push along (i, j) deactivates vertex i and makes $g(i)$ disappear from the sum, whereas $g(j)$ may appear. Furthermore, such a non-saturating push does not create new feasible edges. Furthermore, $g(i) \ge g(j) + 1$, as every vertex reachable from j is reachable from i as well, whereas i may not be reachable from j as the network is acyclic. Thus a non-saturating push decreases the potential function by at least one.

Summarizing:

Theorem 26 *The generic implementation of the cost scaling algorithm has a running time of $O(|V|^2|A|\log(|V|C))$.*

Proof Our bound on the runtime of the inner while-loop is dominated by the non-saturating pushes. This loop is executed $\lceil\log(|V|C + 1)\rceil$ times.   □

Exits

Before we start the next chapter some concluding remarks. Again it is worthwhile to examine vertices. Furthermore, it is rewarded to process the vertices in *topological order*. This is possible since the feasible network is acyclic which enables us to consider the edges as generators of a *partial order* ($u \le v \Leftrightarrow$ there exists a directed u-v-path). A topological order is defined as a total order of the vertices, that is compatible with the partial order, i.e. it contains all its relations. This reduces the bound to $O(|V|^3 \log(|V|C))$. For details we refer to the book of Ahuja, Magnanti und Orlin [1]. For a more recent overview over results and algorithms on minimum cost flows we refer to [28, 39].

Exercises

Exercise 55 A function $c : K \to \mathbb{R}$ from a convex set $K \subseteq \mathbb{R}^n$ to the reals is called a *convex function* if

$$\forall K \ni x, y \in K \; \forall t \in (0, 1) : f(tx + (1 - t)y) \leq tf(x) + (1 - t)f(z).$$

We consider a minimum cost flow problem which is the same as Problem 10 except that the linear cost function $c^\top f = \sum_{(i,j)inA} c_{ij} f_{ij}$ is replaced by a *separable, convex cost function* i.e. a sum of convex cost functions for the flow on each arc

$$c(f) := \sum_{(i,j)inA} C_{ij}(f_{ij}),$$

where all $C_{i,j} : \mathbb{R} \to \mathbb{R}$ are convex functions, additionally requiring the flow on each arc to be a non-negative integer. By adding a constant for each arc we may assume that $C_{ij}(0) = 0$ for all $(i, j) \in A$.

Let f be a feasible flow and define a cost c_r function on the residual network as follows

$$c_r((i, j)) := \begin{cases} C_{ij}(f_{ij} + 1) - C_{ij}(f_{ij}) & \text{if } (i, j) \in A \text{ and } f_{ij} \leq w_{ij} \\ C_{ij}(f_{ij} - 1) - C_{ij}(f_{ij}) & \text{if } (j, i) \in A \text{ and } f_{ij} \geq 1. \end{cases}$$

Prove that f is a flow of minimum cost if and only if the residual network with cost function c_r does nor have a directed circuit of negative weight.

Exercise 56 Model Application 5 as a minimum cost flow problem.

Chapter 8
Matching

Undirected graphs can also be interpreted in another way, namely as a representation of *symmetric binary relations*. Vertices represent objects and an edge $\{u, v\}$ indicates that u and v are related. Take the assignment of students which share a room in a dormitory as an example. The students are our objects and the relation is *sharing-a-room*. The endpoints of any edge are roommates. If there are only double rooms in the dormitory, a pairing of roommates is a *matching*.

Definition 23 Let $G = (V, E)$ be a graph. A subgraph $M = (V, E_M)$ of G is called a *matching*, if each vertex is incident with at most one edge in M; that is $\forall v \in V : d_M(v) \leq 1$. If $(u, v) \in E_M$, we say u is matched to v.

There are many other examples of such assignments, for example lecture halls to university courses, students to classes, or as in the classical marriage problem below, men to women.

8.1 Bipartite Matching

We distinguish two types of matching problems, those which are "two-sided" and those which are not. The latter examples above are all two-sided since we will not match lecture halls to lecture halls or students to students.

The two-sided case is considerably easier as there are two classes of vertices with no adjacency inside each class.

Definition 24 A graph $G = (U, E)$ is called *bipartite*, if there exists a partition $U = V_1 \dot\cup V_2$ of U, such that each edge has exactly one end node in V_1 and the other in V_2.

A famous example problem of bipartite matching is the "marriage problem".

Application 57 In a village there are m women and n men of marriageable age. Each woman discloses a list of acceptable male partners for marriage. Find a matching that pairs off as many men and women as possible, taking their preferences into account while respecting the ban on polygamy.

W. Hochstättler, A. Schliep, *CATBox*, DOI 10.1007/978-3-642-03822-8_8,
© Springer-Verlag Berlin Heidelberg 2010

The bipartite matching problem can easily be transformed into a Max-flow-problem (see Chap. 6) and, indeed, the fastest known algorithm exploits this fact.

Let $G = (V_1 \dot\cup V_2, E)$ be a bipartite graph and consider the network (see Fig. 8.1)

$$N(G) = (\{s\} \cup V_1 \cup V_2 \cup \{t\}, A)$$

with arc set $A = A_1 \cup A_2 \cup A_3$ where

$$A_1 = \{(s, v) \mid v \in V_1\},$$
$$A_2 = \{(v_1, v_2) \mid v_1 \in V_1, v_2 \in V_2, \{v_1, v_2\} \in E\},$$
$$A_3 = \{(v, t) \mid v \in V_2\}.$$

and the capacity function

$$cap\,(a) = \begin{cases} 1 & \text{if } a \in A_1 \\ \infty & \text{if } a \in A_2 \\ 1 & \text{if } a \in A_3. \end{cases}$$

The capacity function is designed so that for any feasible flow in this network at most one unit of flow can enter a vertex of V_1 or leave a vertex of V_2. In an integer flow a unit of flow that enters a vertex of V_1 has to be passed along an edge of G and leave the corresponding vertex of V_2. Hence, a maximal integer flow corresponds to a matching with as many edges as possible. In the following theorem we make this more precise.

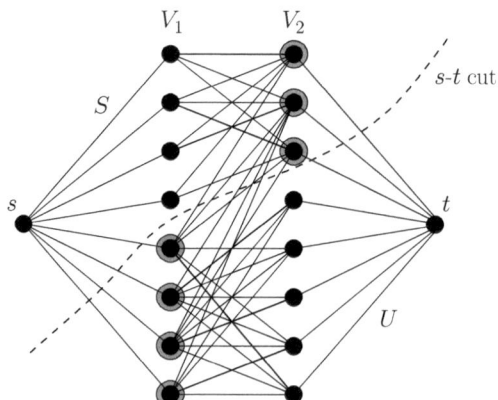

Fig. 8.1 The network in the proof of König's theorem from Max-flow-Min-cut. Vertices with a grey outline belong to C

Theorem 27 *Let* $f : A \to \mathbb{Z}$ *be an integer (maximal) flow in* $N(G)$ *and* M *the set of edges of* G *which correspond to arcs in* A_2 *that have a non-zero flow in* f. *Then*

M is a (maximal) matching. If on the other hand $\widetilde{M}$ is a (maximal) matching in G, then

$$\widetilde{f}_a := \begin{cases} d_M(v) & \text{if } a = (s, v) \text{ or } a = (v, t) \\ 1 & \text{if } a = (v_1, v_2) \text{ and } \{v_1, v_2\} \in \widetilde{M} \\ 0 & \text{if } a = (v_1, v_2) \text{ and } \{v_1, v_2\} \notin \widetilde{M} \end{cases}$$

yields a (maximal) flow in $N(G)$.

Proof Let f be an integer flow in $N(G)$ and M as above. The flow conservation rule implies for all $v_1 \in V_1$, $d_M(v_1) = f_{(s,v_1)} \leq 1$ and also $d_M(v_2) \leq 1$ for all $v_2 \in V_2$. Hence M forms a matching. Furthermore, $|M| = |f|$.

On the other hand, given a matching, the above construction yields a flow $\widetilde{f}$ satisfying $|\widetilde{f}| = |\widetilde{M}|$ for any given matching $\widetilde{M}$ and the claim follows. $\square$

Remark 11 The Max-flow-Min-cut Theorem in the special case of bipartite matching is an older result due to Dénes König. In a bipartite graph the size of a maximum matching equals the minimum size of a vertex cover, that is a vertex set meeting every edge.

Definition 25 Let $G = (V, E)$ be a graph and $C \subseteq V$. Then C is a *vertex cover*, if each edge $e \in E$ has at least one end node in C.

For general graphs the minimum vertex cover problem is one of the classical $\mathcal{NP}$-complete problems and thus it is unlikely that there is a polynomial time algorithm to solve it. For bipartite graphs it is easy.

Theorem 28 (König 1931) *Let $G = (U = V_1 \dot{\cup} V_2, E)$ be a bipartite graph. Then*

$$\max\{|M| \mid M \text{ is a matching of } G\} = \min\{|C| \mid C \text{ is a vertex cover}\}.$$

Proof Let M be an arbitrary matching and C be a vertex cover. Clearly, each edge of M is covered by at least one vertex from C and no vertex of C can cover more than one edge from M, thus $|M| \leq |C|$. Now, let M be a maximal matching and f the corresponding flow as constructed in Theorem 27. By Theorem 15 there must be an s-t-cut $(S, U \setminus S)$ of weight $cap(S) = |f| = |M| < \infty$. In particular $(S, U \setminus S)$ cannot contain an arc from A_2, i.e. a forward arc corresponding to an edge of our bipartite graph.

We set $C = (V_1 \setminus S) \cup (S \cap V_2)$ and claim that C is a vertex cover. Assume to the contrary that there is an arc from $V_1 \cap S$ to $V_2 \setminus S$. This arc is directed from S to $U \setminus S$ and, thus is an arc of infinite capacity in $(S, U \setminus S)$, a contradiction.

Finally, we show that $|C| \leq cap(S) = |f| = |M|$, proving the other inequality. Let $v \in C$. If $v \in V_1$ then $(s, v) \in (S, U \setminus S)$, otherwise $(v, t) \in (S, U \setminus S)$. In any case we have $|C| \leq cap(S)$. $\square$

Matching algorithms work similar to the way the Ford-Fulkerson algorithm from Sect. 6.2 augments a flow, and usually augment a matching until it is maximal. The fastest known method for bipartite matching is a hybrid method from two augmentation strategies for flow problems and runs in $O(|E|\sqrt{|U|})$ (see [1] for details). We present a simpler version of an augmentation algorithm that can directly be considered as a specialization of the Edmonds-Karp implementation of the algorithm of Ford and Fulkerson which we presented in Sect. 6.2. Therefore the main loops are very similar.

ALGORITHM Bipartite

```
42  maximal = False
    while not maximal:
        P = FindAugmentingPath(V1,V2,M)
        if P:
            Augment(P,M)
47      else:
            maximal = True

    ShowVertexCover(V1,V2)
```

Software Exercise 58 Run the algorithm `Bipartite.alg` on the graph `Koenig.cat` and compare the dynamics to the behavior of Ford-Fulkerson on the network `Koenig.cat`.

We can organize the search for an augmenting path without adding an artificial source and sink. The *matching edges* that support a flow correspond to the edges that point from V_2 to V_1 in the residual network. In the following we will call an unmatched vertex *exposed*. Starting with an exposed vertex from V_1 we try to find a path that alternates between non-matching and matching edges to an exposed vertex from V_2. We will examine these *alternating paths* in more detail in the next section.

ALGORITHM Bipartite

```
    def FindExposedVertex(V,M):
        DeleteLabels()
        for v in V.Vertices():
4           if not v in M.Vertices():
                pred[v] = v
                Q.Append(v)

        while Q.IsNotEmpty():
9           v = Q.Top()
            for w in Neighborhood(v):
                if not pred[w]:
                    pred[w] = v
                    if w in M.Vertices():
14                      u = MatchingPartner(w)
                        pred[u] = w
                        Q.Append(u)
                    else:
                        return w
19      return None
```

Software Exercise 59 Most of the subroutines in `Bipartite.alg` are traceable. When you start the algorithm you should step to `P=FindAugmentingPath` `(V,W,M)` and then trace into this subroutine as well as into `FindExposedVertex` `InW(V,W,M)`.

First, we delete all old labels and collect all unmatched vertices in the first *color class V* in a queue Q. This queue will always consist solely of vertices from V. We say that these vertices have an odd label. Starting with the top element v=Q.Top() of that queue we label all unlabeled neighbors w by making v=pred[w] their predecessor, they receive even labels. If such a w is unmatched, we have found an augmenting path, otherwise we append the matching partner of w to the queue and attach an odd label to it.

This search yields a forest, where the root of each component is an unmatched vertex from V. To observe this, try the graph Forest1.cat and run the matching algorithm on it until edge (6, 12) becomes matched. Now, tracing will yield three search trees.

Each time an exposed vertex w is detected, we construct the augmenting path P by following the pred-pointers and the matching edges. Taking the symmetric difference of P and the matching yields a larger matching.

ALGORITHM Bipartite

```
21 def FindAugmentingPath(V1,V2,M):
       P = []
       w = FindExposedVertex(V1,M)
       if w:
           while pred[w] != w:
26             P.append((pred[w],w))
               w = pred[w]
       return ShowPath(P)
```

Now we can prove

Lemma 26   *(i) At any stage of the algorithm M is a matching of G.*
(ii) The augmentation step in function Augment(P,M) *increases the size of the matching by 1.*
*(iii) The **while** loop in line 43 is executed at most* $|V| + 1$ *times.*
(iv) When the algorithm terminates because maximal = True, *then the unlabeled vertices of* V1 *and the vertices with (even) label in* V2 *form a minimal vertex cover C and* $|M| = |C|$.

Proof   The first statement is clearly correct at the beginning of the algorithm. The matching is changed only in the augmentation step in Augment(P,M). By construction, the path P starts in a vertex $v_1 \in V_1$ and ends in a vertex $v_k \in V_2$. Furthermore, such a path starts and ends with a non-matching edge and alternates between matching edges and non-matching edges. In particular $|P \setminus M| = |M \cap P| + 1$. Therefore, the adjacencies of a vertex v in M change only if v is either an unmatched vertex and P starts or ends in v or if v is a matched vertex that is traversed by P. In the latter case one edge is added to v and one is deleted. In any case we maintain a degree of at most one in each vertex, implying the first two statements.

In the while loop maximal remains false only if an alternating path has been found, implying the third assertion.

If maximal has been set to true, then no alternating path has been found. Thus, we have found a set $S \subseteq V_1$ of vertices such that all their neighbors have a smaller label. Moreover, no labeled vertex $v \in V_1$ is adjacent to an unlabeled vertex $w \in V_2$. Thus the unlabeled vertices in V_1 together with the labeled vertices in V_2 form a vertex cover. By definition, all of these vertices are matched and since the matching partner of a vertex with an (even) label must be labeled as well, no two of them are matched to one another. Thus each matching edge is covered exactly once, implying $|C| \leq |M|$ and thus the assertion.   □

Theorem 29 *The algorithm computes a maximal matching in* $O(|V|\,|E|)$.

Proof Correctness is implied by the last lemma. There are at most $|V|$ augmentation steps each of which consists of a breadth first search in a directed graph (matching edges are directed backwards). If $|E| < |V|$ we have at most $|E|$ augmentations. Thus in any case we are done in $O(|V|\,|E|)$.   □

We end this section with a classical application of König's Theorem.

Definition 26 Let S be a finite set and $(Q_i)_{i \in I} \subseteq 2^S$ a family of subsets of S with index set I. A *system of distinct representatives (SDR)* of $(Q_i)_{i \in I}$ is an injective map $f : I \to S$ such that $\forall i \in I : f(i) \in Q_i$. We call the image of such a map a *transversal*. If such a map is only partially defined, i.e $J \subseteq I$ and $g : J \to S$ is injective, the $g(J)$ is a partial transversal.

Theorem 30 (Hall 1935) *The family* $(Q_i)_{i \in I}$ *has a SDR or a transversal if, and only if,*

$$\forall H \subseteq I : |\bigcup_{i \in H} Q_i| \geq |H|.$$

Proof If the family has a transversal, then the condition, clearly, must be satisfied. In order to prove the other implication, we consider the bipartite graph $G = (I \dot\cup S, E)$ defined by $(i, s) \in E \Leftrightarrow s \in Q_i$. Then, transversals correspond to the matchings of cardinality $|I|$. By König's Theorem the family has no transversal if and only if the graph has a vertex cover C where $|C| < |I|$. Let $H = I \setminus C$. The edges incident with H then have to be covered by $T := C \cap S$, hence $N(H) \subseteq T$, where $N(H)$ denotes the vertices adjacent to H. Furthermore,

$$|I| = |H| + |C \cap I| > |C| = |C \cap I| + |T| \geq |C \cap I| + |N(H)|$$

and thus

$$|H| > N(H) = |\bigcup_{i \in H} Q_i|.$$

□

We call a matching M *perfect* if $d_M(v) = 1$ for all $v \in V$. The following theorem is as well a predecessor and a special case of Hall's Theorem.

Corollary 8 (Marriage Theorem of Frobenius 1917) *A bipartite graph* $G = (V_1 \dot\cup V_2, E)$ *has a perfect matching, if and only if,* $|V_1| = |V_2|$ *and* $|H| \leq |N(H)|$ *for all* $H \subseteq V_1$.

Exercise 60 A graph $G = (V, E)$ is called *d-regular* for $d \in \mathbb{Z}_+$ if $d_G(v) = d$ for all $v \in V$. Let $G = (V_1 \dot\cup V_2, E)$ be a bipartite d-regular graph. Show that

  (i) $|V_1| = |V_2|$,
 (ii) G has a perfect matching,
(iii) E is the disjoint union of d perfect matchings.

Exercise 61 Let $D = (V, A)$ be a digraph. We construct a corresponding bipartite graph G as follows. Let V' and V'' be disjoint copies of V. To each arc $(u, v) \in A$ we construct an edge connecting the vertex $u' \in V'$ to $v'' \in V''$ where u' and v'' are the vertices corresponding to u respectively v. Show that

  (i) A matching in G corresponds to a disjoint union of directed circuits and directed paths in D.
 (ii) If D is acyclic, which means that D has no directed circuit, then a maximum matching in G corresponds to a minimum path cover in D, more precisely, if the cardinality of a maximum matching in G is ν, then the size of a minimum path cover in D is $|V| - \nu$.
(iii) Deduce the Theorem of Dilworth: The size of a minimum path cover in a directed acyclic graph G equals the maximum cardinality of an antichain, i.e. a set of vertices that are pairwise not connected by a directed path.

Exercise 62 A real-valued matrix $A = (a_{ij})$ is called doubly-stochastic, if all entries are in the interval $[0, 1]$ and all columns as well as rows sum to 1. A doubly stochastic matrix where all entries are either 0 or 1 is called a permutation matrix. Show that any doubly stochastic matrix is a convex combination of permutation matrices.

8.2 Augmenting Paths

Although we did not state it explicitly, the correctness of the Ford-Fulkerson algorithm implies that a flow is maximal if, and only if, it does not admit an augmenting path. In the case of bipartite matching such an augmenting path alternates between edges not in M and matching edges. And this concept generalizes to the non-bipartite case.

Definition 27 Let $G = (V, E)$ be a (not necessarily bipartite) graph and M a matching in G. A path $v_1, \ldots, v_{2k}$ is called M-augmenting, if

$$\deg_M(v_1) = \deg_M(v_{2k}) = 0 \text{ and } (v_{2i}, v_{2i+1}) \in M \text{ for all } i = 1, \ldots, k - 1.$$

If S_1, S_2 are sets, then we define the *symmetric difference* $S_1 \triangle S_2$ as

$$S_1 \triangle S_2 := (S_1 \cup S_2) \setminus (S_1 \cap S_2).$$

Thus, an M-augmenting path starts and ends in non-matched vertices and alternates between matching edges and non-matching edges. It can be utilized to increase the size of a given matching (see Fig. 8.2).

For bipartite graphs we already know about the following theorem, but it also holds in non-bipartite graphs.

Theorem 31 *Let $G = (V, E)$ be a (not necessarily bipartite) graph and $M = (V, F)$ a matching in G. Then M is of maximal cardinality if and only if there is no M-augmenting path.*

Proof Let $v_0, \ldots, v_{2k}$ be an M-augmenting path and P its set of edges. We set $\widetilde{F} := F \Delta P$. Then $\widetilde{M} = (V, \widetilde{F})$ is a matching and $|\widetilde{F}| = |F| + 1$.

On the other hand let $\widetilde{M} = (V, \widetilde{F})$ be a matching and $|\widetilde{F}| > |F|$. We set $A = \widetilde{F} \Delta F$. All vertices of the graph $H = (V, A)$ have a degree from $\{0, 1, 2\}$. Therefore H consists of paths and even circuits that alternate in F and $\widetilde{F}$. Since $|\widetilde{F}| > |F|$ one of these paths has to consist of more edges from $\widetilde{F}$ than from F. Thus, an M-augmenting path has been encountered. $\square$

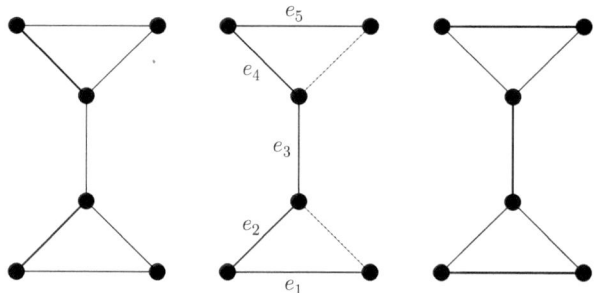

Fig. 8.2 The matching $\{e_2.e_4\}$ is augmented by the path $e_1e_2e_3e_4e_5$ to the matching $\{e_1, e_3, e_5\}$

8.3 Non-Bipartite Graphs

Our results about augmenting paths suggest that we proceed as in the bipartite case. As long as we find an augmenting path we use it to augment the matching. Otherwise, we are done.

But, of course, the devil is in the details, namely in the question of how can we detect an augmenting path? Recall how we proceed in the bipartite case: we grow a search forest rooted in the unmatched vertices of one color class until we reach an unmatched vertex from the other color class. How could we possibly extend this method to the non-bipartite case?

If we number the vertices of an augmenting path, its end vertices will receive numbers of different parity. We can take this for granted in the bipartite case since all vertices of the same parity belong to the same color class. The approach we will follow here is to try to divide the vertices into odd and even vertices. This will be compatible with our approach in the bipartite case, since bipartite graphs do not have odd circuits. In the presence of odd circuits a simple labeling procedure may fail to find an augmenting path. We show a small example demonstrating the possible problems with odd circuits in Fig. 8.3.

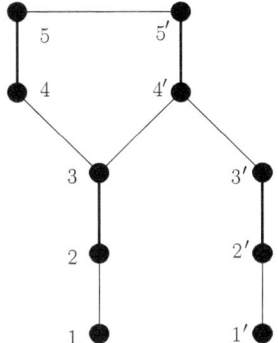

Fig. 8.3 We do not know how to easily generalize the search procedure from the bipartite case to this situation

Our labeling strategy distinguishes between *odd (outer)* and *even (inner)* vertices. If we grow a search forest rooted at all unmatched vertices as before, both unmatched vertices will receive an odd label. We sketched the result of such a labeling strategy in the figure. If we alternatively grow a search tree from either one of the unmatched vertices, the other one will also receive a label of the same parity. Note, in both cases we overlook the alternating path labeled $1, 2, 3, 4, 5, 5', 4', 3', 2', 1'$. The reason is that there is no natural partition into odd and even vertices such that no two odd respectively even vertices are adjacent.

Exercise 63 A graph $G = (V, E)$ is bipartite if and only if it does not contain a circuit of odd length.

To extend our algorithm we therefore have to overcome the problem with certain odd circuits. Which odd circuits do we refer to? The odd circuits that cause trouble in our search procedure are those which are almost perfectly internally matched except for *one* vertex.

In general, any odd circuit C has the following interesting property.

Definition 28 Let $G = (V, E)$ be a graph. We say that G is *hypomatchable* if for all $v \in V$ the graph $G \setminus \{v\}$ has a perfect matching

Deleting any vertex from an odd circuit leaves a path with an odd number of edges, and thus odd circuits are hypomatchable.

Now consider the case that we are given a matching $M = (V, F)$ and an odd circuit C of length $2k + 1$, such that

$$|M \cap C| = k.$$

Note that exactly one vertex of C is exposed i.e unmatched. However, we can make any vertex in C exposed by modifying the matching along the odd circuit C. Therefore, we can simplify the problem by treating an odd circuit, which we will call a *blossom* in the following, just like a single unmatched vertex and shrink it to a *super node*. Iterating this process yields the following recursive definition.

Definition 29 Let $G = (V, E)$ be a graph. A vertex set $B \subseteq V$ of G is called a *blossom* if B contains the vertex set $V(C)$ of an odd circuit C of G with edge set $E(C)$ such that either $B = V(C)$ or $B/E(C)$ is a blossom.

Definition 30 Let $G = (V, E)$ be a graph, $B \subseteq V$ a set of vertices and $F \subseteq E$ a set of edges such that $T = (B, F)$ is a tree. The vertex b we obtain from the vertex set B by contracting all edges in F is called a *super node*. The graph that results if we additionally remove all loops and parallel edges produced by the contractions is said to be obtained by shrinking B.

Proposition 8 *Let V be a blossom in the graph $G = (V, E)$. Then G is hypomatchable.*

Proof We proceed by induction on the number of recursion steps in the definition. Odd circuits are obviously hypomatchable. Thus let C be an odd circuit in G such that G/C is a blossom and $v \in V$. If $v \notin V(C)$, G/C has a matching M that matches all vertices but v by induction. Now, M is a matching in G that leaves all but one vertex of C unmatched. Since C is hypomatchable, the claim follows.

If on the other hand $v \in V(C)$ then G/C has a matching that leaves the super node unmatched. Again the claim follows since C is hypomatchable. $\square$

Extending the bipartite matching algorithm by the shrinking of blossoms yields the following sketch of an algorithm:

   (i) start an alternating search;
  (ii) when a blossom is detected, shrink it;
 (iii) when an augmenting path is found in the shrunken graph, expand the blossoms of the super nodes which lie on the path, complete it to an augmenting path in the original graph, augment the matching and start again.

Our main loop thus is as follows

ALGORITHM CardinalityMatching

```
while not maximal:
    (e, pred, base, label) = FindSuspectEdge(SG, SGM, SGA)
    if e is None:
        maximal = True
    else:
        if base[e[0]] == base[e[1]]: # Found a blossom
            B = BlossomContaining(e, pred)
            sv = ShrinkBlossom(B)
        else:
            P = FindAugmentingPath(SG, SGM, pred, e)
            Augment(SGM, P)

ShowGallaiEdmondsDecomposition(label)
```

The subroutine `FindSuspectEdge` returns an edge e that connects two outer (odd) vertices. Furthermore it returns three arrays associated with the vertices,

– the predecessor `pred` from which the vertex has been reached,
– the `base` indicating the root of the search tree, and
– the `label` which tells us, whether the vertex is inner, outer, or has not been found yet.

If the bases of the end vertices of edge e coincide, a blossom has been detected. We shrink it and continue. Otherwise, we have found an augmenting path that we use to augment the matching. If we do not find a `SuspectEdge` the current matching is maximal. As in the case of bipartite matchings the algorithm also constructs a short certificate for optimality. It is a bit more complicated than a vertex cover and is called the *Gallai-Edmonds-Decomposition*. We will address it in greater detail later on.

Let us discuss our strategy to detect blossoms or alternating paths in more detail. This is done by classifying the vertices as inner (even) and outer (odd). Initially, we label all exposed vertices outer and put them into the queue, which will always contain only outer (odd) vertices.

We process an outer vertex from the queue as in the bipartite case, as long as all neighbors are either unlabeled or inner vertices. Note that, given an augmenting path, when we label both ends as outer and then label the neighbors inner and their matching partners outer, necessarily some non-matching edge has to have two outer end vertices. The same holds for blossoms. Here we have an odd alternating circuit where the unmatched vertex is outer. If we cut the circuit at that vertex which we duplicate, we get a path where the above argument applies.

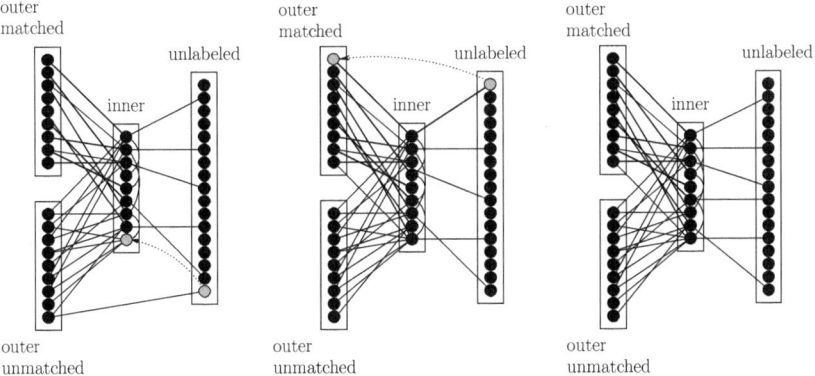

Fig. 8.4 Three possible steps in the labeling procedure. In the *leftmost* figure we find an unlabeled vertex adjacent to an outer vertex. If this new vertex is unmatched, an augmenting path has been detected. Otherwise, we make its matching partner, which necessarily is unlabeled, an outer vertex and add it to the queue. This is depicted in the *middle*. Finally, when the labeling terminates, we find that no unlabeled vertex is adjacent to any of the outer vertices and no pair of outer vertices is connected by an edge. Furthermore, all unlabeled vertices are internally matched

Therefore we search for an edge where both end vertices are outer. In order to distinguish between augmenting paths and blossoms, we propagate the roots of the search trees to their descendants. This is done in the array `base`.

Summarizing, our search trees are grown by examining outer vertices, that is vertices with odd labels. We remove an outer vertex z from the queue and consider all its neighbors that do not already have received an inner (even) label. We label all unlabeled vertices among them as inner (even) and their matching partners, which must exist, because all exposed vertices are outer, as outer and append them to the queue. Finally, if one of z's neighbors, say x has already received an outer (odd) label one of the following must occur:

(i) If we find that such an outer (odd) vertex z belonging to the tree with root (base) v_0 has an outer neighbor x belonging to a different tree rooted at w_0, then we find an augmenting path from v_0 to w_0 if we connect the unique $v_0 z$-path and the $x w_0$-path by the edge (xz).

(ii) If we detect an odd neighbor w in the same component, we find a blossom by backtracking to the first common node of the $z v_0$ and the $w v_0$ path.

We implemented this as follows:

ALGORITHM CardinalityMatching

```
def FindSuspectEdge(G, M, A):
    A.Clear()
    Q = TrackingVertexQueue(A)
    pred = SearchTreePredecessorLabeling(A)
    label = VisibleVertexLabeling(A)
    base = {}

    for v in UnmatchedVertices(G, M):
        pred[v] = v
        base[v] = v
        label[v] = 1
        Q.Append(v)

    while Q.IsNotEmpty():
        v = Q.Top()
        for w in G.Neighborhood(v):
            if not pred.QDefined(w):
                pred[w] = v
                base[w] = base[v]
                label[w] = 0

                if w in M.Vertices():
                    u = M.Neighborhood(w)[0]
                    pred[u] = w
                    base[u] = base[w]
                    label[u] = 1
                    Q.Append(u)
            elif pred[v] != w and label[w] != 0:
                A.SetEdgeColor(v,w,'red')
                return ((v,w), pred, base, label)
    return (None, pred, base, label)
```

By our previous discussion the function `FindSuspectEdge` returns an edge with both end vertices labeled odd, leading to either a blossom or an augmenting path. If no such edge is found, `maximal` is set to `True`. We delay the discussion of that case. With some effort we will derive a duality theorem from that situation later on. This will give an alternative proof of correctness of the algorithm. First we show that the "reduction to the bipartite case" works properly. For that we note the following:

Lemma 27   *(i)   Whenever a blossom is found, its* bud, *i.e. the vertex where the paths meet, is outer and the super node that the blossom is shrunk into will receive the same (odd) label if we start a search in the shrunken graph. Thus a super node will never be an inner vertex.*
 (ii)   All vertices that are exposed are outer vertices, in particular all vertices with an even label are matched.
(iii)   When Maximal is set True any outer vertex is adjacent only to inner vertices.

Proof   The second statement is trivial and the first follows from the fact that inner vertices have only one successor, their matching partner. If an outer vertex is found to be adjacent to another outer vertex, then either a blossom or an augmenting path is found. When the routine `FindSuspectEdge` returns `None`, the neighborhood of all outer vertices has been scanned and given a label.   □

Consequently, when `maximal` is set `True`, if we remove the arcs connecting vertices with even label from the shrunken graph G', we obtain a bipartite graph G'' on the set of vertices that did receive a label. Furthermore, the tree search we performed on G' is identical to applying the labeling algorithm to G''. Clearly, a maximal matching will not match two inner vertices, since all of them are matched to an outer vertex, all exposed vertices are outer, but no two outer vertices are adjacent. Therefore, the correctness of the bipartite matching algorithm implies, that the present matching in G' is maximal.

The correctness of the whole algorithm now follows, if we can prove that shrinking a blossom does not affect the existence of an augmenting path.

Theorem 32 *Let M be a matching in $G = (V, E)$ and $B \subseteq V$ be a blossom that has been detected by the procedure* `FindSuspectEdge()` *and let E_B denote the edges induced by B in G. There exists an M-augmenting path in G if and only it there exists an $(M \setminus E_B)$-augmenting path in G/E_B.*

Proof   Obviously, we can always extend an augmenting path in G/E_B to one in G, since the blossoms are hypomatchable (see Fig. 8.5. We thus have to show that the existence of an augmenting path in G implies the existence of an augmenting path in G/E_B, this is shown in the following Lemma.   □

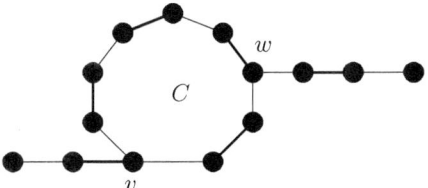

Fig. 8.5 Extending an augmenting path in an odd circuit

Lemma 28 *Let $G = (V, E)$ be a graph and M a matching. Let B, E_B be as in Theorem 32. Let W_{v_s, v_0} be the alternating path constructed in the procedure* `FindSuspectEdge()` *that led to the detection of B. Thus, W_{v_s, v_0} starts in an exposed vertex v_s and leads to the unique (outer) vertex $v_0 \in V_B$ that is not matched to any vertex of V_B. Assume $\overline{M}$ is a matching of G such that $|\overline{M}| > |M|$, then there exists a matching $\widetilde{M}$ of $G' := G/E_B$ such that $|\widetilde{M}| > |M \setminus E_B|$. In particular there exists an M-augmenting path in G if and only if there exists an $M_{|G/B}$-augmenting path in G/B.*

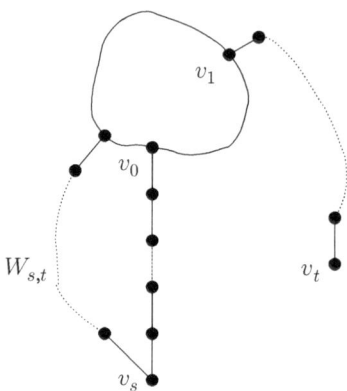

Fig. 8.6 Shifted matching and $s = s'$

Proof First we shift the matching M to get v_0 exposed, as follows. Consider the matching $\widehat{M} := M \triangle W_{v_s, v_0}$. Then $|M| = |\widehat{M}|$ and v_0 is unmatched in $\widehat{M}$. Since $|\overline{M}| > |\widehat{M}|$ the symmetric difference $\overline{M} \triangle \widehat{M}$ contains an $\widehat{M}$-augmenting path $W_{v_{s'}, v_t}$. If this path does not meet B, it remains an $\widehat{M}$-augmenting path in G/E_B. Otherwise, as v_0 is exposed, and all other vertices of V_B are matched inside of B, $W_{v_{s'}, v_t}$ either has to start in v_0 or to enter and leave B via non-matching edges. Let v_1 denote the last vertex on $W_{s', t}$ that belongs to V_B, and W_{v_1, v_t} the v_1-v_t-segment of this path (see Fig. 8.6) leading to an exposed vertex $v_t \notin V_B$. Clearly, W_{v_1, v_t} is a $\widehat{M} \setminus (E_B)$-augmenting path in G/E_B. As $|\widehat{M}_{|G/B}| = |M_{|G/B}|$, the claim follows.   □

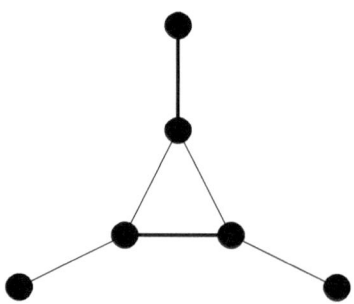

Fig. 8.7 The stem is necessary

Actually, in Lemma 28 it is sufficient to require that there is an alternating path from an exposed vertex to the blossom that ends in a matching edge. If there is no alternating path from an exposed vertex to the bud of the blossom, the claim is false, see Fig. 8.7.

We end this section with a possible implementation of the hypomatching of a blossom B. We organize its recursive tree structure as follows. Each time we shrink an odd circuit C we create a new (super) node c that has a pointer `blossom[supervertex[c]]` to an ordered list of the vertices of C.

When we encounter an augmenting path P that passes through a blossom B corresponding to super node b in $G/E(B)$, `blossom[supervertex[b]]` yields an odd circuit C. Let v respectively w be the unique vertices in C that are adjacent to a matching edge respectively to a non-matching edge in P in the graph where C is expanded. If $v = w$ we are done. Otherwise w is matched in C and v is not. The matching edge at w provides an orientation for C and we shift the matching in C along that direction until v becomes matched. If v is a super node we apply this procedure recursively, otherwise we are done.

In our implementation of the prolog of `CardinalityMatching.alg` you will see that we do not explicitly perform that shift until v is met, but rather alternate non-matching and matching edges on all but the last edge of C. Note that this yields the same result.

Exercise 64 Compare the computational complexity of the hypomatching as described above and as implemented in the prolog of `CardinalityMatching. alg`.

8.4 The Tutte-Berge Formula

In this section we will present the promised duality result which also yields an alternative proof of the correctness of Edmonds' blossom algorithm.

Recall the situation when our algorithm terminates. We have outer, inner and unlabeled vertices. No pair of outer vertices is adjacent and, furthermore all super nodes are outer (see Fig. 8.4).

Let us put the set U of unlabeled vertices aside for a while. Since all unmatched vertices are outer, they are perfectly matched. Furthermore, they are neither adjacent to any outer vertex nor matched to any inner vertex.

Ignoring the edges that connect inner vertices, the labeled vertices in the shrunken graph G form a bipartite graph. Thus, by König's Theorem and our analysis of the bipartite case in Lemma 26, the set of inner vertices A, that is the vertices with an even label, form a minimum vertex cover in the shrunken graph G' restricted to the labeled vertices. Note that they, trivially, also cover the edges that connect two inner vertices.

Taking also the unlabeled vertices U into account we find that $G' \setminus A$ consists of isolated outer vertices and perfectly matched unlabeled vertices U. Recall, that super nodes are always outer vertices. Thus, A consists solely of vertices of the original

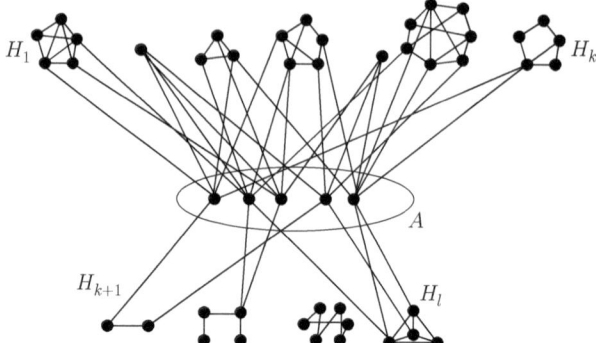

Fig. 8.8 An example for the Tutte-Berge formula

graph G, and we can consider the components of $G \setminus A$. Since super nodes correspond to blossoms in the original graph, the components consist of (see Fig. 8.8)

- perfectly matched connected subgraphs $H_{k+1}, \ldots, H_l$, which have unlabeled vertices. The number of vertices in all these components is even, since they are perfectly *internally* matched, and
- isolated unmatched vertices and hypomatched blossoms $H_1, \ldots, H_k$.

On the other hand, what is the size of our matching? A matching edge either

- is part of a hypomatching of a blossom or
- it is also a matching edge in G', and either both endpoints are labeled and exactly one of them belongs to A, or both endpoints are unlabeled and belong to U.

Thus the size of our matching is

$$|M| = \sum_{i=1}^{k} \frac{|H_i| - 1}{2} + \sum_{i=k+1}^{l} \frac{|H_i|}{2} + |A|$$

$$= |A| + \sum_{i=1}^{l} \left\lfloor \frac{|H_i|}{2} \right\rfloor.$$

The set A is called *barrier*. The above analysis yields the promised duality result. We have the following theorem:

Theorem 33 (Tutte-Berge-Formula)

$$\max\{|M| \mid M \text{ is Matching}\} = \min \left\{ |A| + \sum_{\substack{|H_i| \\ \text{components of } G \setminus A}} \left\lfloor \frac{|H_i|}{2} \right\rfloor \mid A \subseteq V \right\}.$$

Proof By the above analysis there exists a matching and a barrier such that equality holds, therefore the minimum of the right hand side is at most the size of a maximum matching.

Now, let $\widetilde{A} \subseteq V$ be arbitrary and $\widetilde{M}$ be any matching. We partition the matching edges according to the number of endpoints they have in $\widetilde{A}$ into $\widetilde{M} = \widetilde{M}_0 \,\dot\cup\, \widetilde{M}_1 \,\dot\cup\, \widetilde{M}_2$. Since the endpoints of each matching edge in M_0 lie in the same component of $G \setminus \widetilde{A}$ we have

$$|\widetilde{M}_0| \le \sum_{\substack{H_i \\ \text{components of } G \setminus \widetilde{A}}} \left\lfloor \frac{|H_i|}{2} \right\rfloor,$$

and $|\widetilde{A}| \ge |M_1| + 2|M_2|$, implying that the size of any matching is bounded from above by any of the expressions on the right hand side that we take the minimum of. $\square$

As a corollary we derive the following non-bipartite analogue of Frobenius' marriage theorem due to Tutte:

Corollary 9 Let $G = (V, E)$ be a graph. Then G has a perfect matching, if and only if, for all $A \subseteq V$ the number of odd components in $G \setminus A$ is at most $|A|$.

Proof By the Tutte-Berge formula G does not have a perfect matching if, and only if, there exists a barrier A such that

$$\frac{|V|}{2} \quad > \quad |A| + \sum_{\substack{H_i \\ \text{components of } G \setminus A}} \left\lfloor \frac{|H_i|}{2} \right\rfloor$$

$$\Leftrightarrow |V| \quad > \quad 2|A| + \sum_{\substack{H_i \\ \text{components of } G \setminus A}} |H_i| - k$$

where k denotes the number of odd components of $G \setminus A$. Subtracting $|V|$ we derive $|A| - k < 0$. $\square$

Exits

The implementation presented here has a complexity of $O(|V|^2|E|)$. There is room for improvement. The best algorithm known was announced by Micali and Vazirani in 1980 and has a complexity of $O(\sqrt{|V|}|E|)$. In a generalization of matching problems one can as well consider subgraphs which have a specified degree (of at most) $b(v)$ where $b : V \to \mathbb{N}$. These problems are known as b-factor problems (b-matching-problems) and can be reduced to classical matching problems using splitting techniques. Note that this does not yield a polynomial time algorithm if we allow edges with multiplicities, where one requires a slightly different approach.

For a detailed survey about existing algorithms and further information again [39] is the best starting point.

Exercises

Exercise 65 A matching in a graph is called maximal if its edge set is not properly contained in the edge set of another matching. Let M_1, M_2 be two maximal matchings with respect to the same graph. Let $|M_i|$ denote the number of edges of M_i. Show that $|M_1| \leq 2|M_2|$.

Exercise 66 Let $G = (V, E)$ be a connected graph. A vertex $v \in V$ is called a *cut vertex* if $G \setminus v$ is disconnected. Show that any 3-regular graph without a cut vertex has a perfect matching.

Exercise 67 The local TV station of Marathon wants to report on the classical long distance run. There are k locations $p_1, \ldots, p_k$. The first runner is expected to arrive at location p_i at time s_i, and at time t_i the last runner to report about will have passed. It takes a_{ij} to move a camera team from p_i to p_j. The TV station wants to use as few camera teams as possible to broadcast from all k locations p_i from s_i to time t_i. Model the problem to minimize the number of camera teams as a bipartite maximum matching problem.

Chapter 9
Weighted Matching

Historically, the weighted matching algorithms we are going to present are the prototypes of primal-dual algorithms. They follow closely the method introduced in Chap. 4. Let us briefly recall how we proceeded there.

Given a weighted problem, the MST, with a corresponding cardinality problem, the computation of connected components, we found a polyhedral description of the convex closure of the set of characteristic vectors of feasible solutions, i.e. the spanning trees. The non-trivial constraints of that polyhedron were of the form

$$\sum_{e \in \partial(P)} x_e \geq |P| - 1,$$

where P was a partition of the set of vertices. The interpretation of the above inequality is that the edges that have their end points in different classes of the partition—we call them *connecting*—must have enough weight to be able to connect them.

The objective of the corresponding *dual program* is to assign non-negative weights to the partitions, maximizing their sum, such that for each edge $e \in E$ the sum of the weights of those partitions where e connects two of its classes is at most the length of e. That was the meaning of the condition (4.9)

$$u_+ - u_- + \sum_{e \in \partial(P)} u_P \leq c_e.$$

Using complementary slackness from Theorem 5 we have that a spanning tree T and a set of feasible weights for the partitions of the vertex set is a dual pair of optimal solutions if, and only if, for each edge $e \in T$ the dual weights are tight; i.e. the sum of the weights of those partitions for which e connects two of its classes equals the length of e.

Since there are trivial dually feasible weights, namely weight zero everywhere, we have a feasible initial solution for the following procedure: If the graph G of the tight edges of the present dual solution is connected, compute a spanning tree. By complementary slackness this must be an MST. Otherwise, the tight edges induce a non-trivial partition of the set of vertices of G. The number of these components is

W. Hochstättler, A. Schliep, *CATBox*, DOI 10.1007/978-3-642-03822-8_9,

one more than the number of edges in a maximum spanning forest of G. We compute the partition and increase the weight of this partition until the next edge becomes tight. The new dual weights clearly are still feasible. Since we decrease the number of components of G with each dual update, this algorithm will terminate using at most $|V| - 1$ dual updates.

As mentioned above, the weighted matching algorithms are prototypes of these primal dual algorithms. In order to apply the technique here, we have to find polyhedral descriptions of the sets of matchings, such that the dual variables have an interpretation in terms of a duality theorem for a matching cardinality problem.

If we consider the complete graph induced by points in the plane, where the weights of the edges are given by the Euclidean distance, the weighted non-bipartite matching algorithm has a pretty geometric interpretation, see Fig. 9.1, presented first by M. Jünger and W.R. Pulleyblank [26]. CATBox contains an implementation of that interpretation only for the non-bipartite case. Nonetheless, the algorithm and the theory for the bipartite case, which we will consider in the next section, are considerably easier and hopefully help to grasp the ideas for the non-bipartite case.

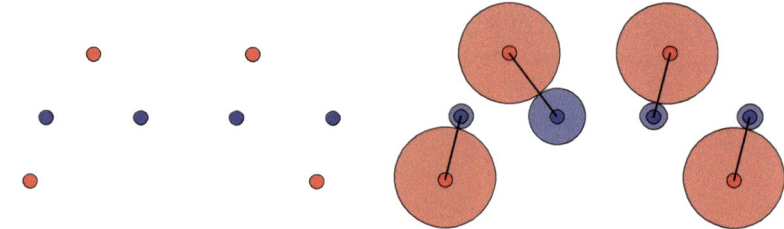

Fig. 9.1 Geometric weighted matching duality

9.1 Bipartite Weighted Matching

Given a graph $G = (V, E)$ and a weight function $w : E \to \mathbb{R}$, the objective of this chapter is to find a maximum or minimum weight matching respectively a minimum or maximum weight *perfect* matching. Maximization and minimization are equivalent under multiplication of the edge weights with -1. Here, we can make our life a lot easier by reducing these four cases to minimizing perfect matchings in *complete graphs* either the *complete bipartite graph* K_{nn} in the bipartite or K_n in the non-bipartite case:

Definition 31 The bipartite graph $K_{nm} = (V, E)$, where $V = V_1 \dot\cup V_2$, $|V_1| = n, |V_2| = m$ and $E = \{(v_1, v_2) \mid v_1 \in V_1, v_2 \in V_2\}$ is called *complete bipartite graph*. The graph $K_n = (V, E)$, where $E = \{(v_1, v_2) \mid v_1, v_2 \in V\}$ is called *complete graph*.

If we search for a minimum weight matching, we will not consider edges of positive weight, as we do not require the matching to be perfect, thus we can delete those edges. We can always assume that the underlying graph is complete, and moreover that $n = m$ in the bipartite case and that n is even in the non-bipartite case. If this does not hold for our input we can simply add one (dummy) vertex, respectively $|n - m|$ dummy vertices in the bipartite case, and all the missing edges with weight zero to guarantee the existence of a perfect matching.

We can safely add a constant c to all edge weights such that all weights become non-negative, as the weight of all perfect matchings for a given graph changes by the same amount. Therefore, it suffices to consider the following problem, which we formulate for the bipartite case first.

Problem 11 Let K_{nn} be the complete bipartite graph and $c : \{1, \ldots, n\}^2 \to \mathbb{Z}_+$ a non-negative, integer weight function on its edges. Find a perfect matching M, such that $\sum_{e \in M} c_e$ is minimal.

Exercise 68 Recall the problem of determining a shortest path. Why is it infeasible to add a constant c to all edge weights such that all weights become non-negative in that case?

Now, the first task is to model the convex hull of the incidence vectors of perfect matchings (recall Definition 8) as a polyhedron in $\mathbb{R}^E$. Since in a perfect matching each vertex is adjacent to exactly one edge, we immediately arrive at the conditions

$$\sum_{j=1}^{n} x_{ij} = 1 \text{ for each } i = 1, \ldots, n \qquad \text{and} \sum_{i=1}^{n} x_{ij} = 1 \text{ for each } j = 1, \ldots, n,$$

where x_{ij} denotes the variable corresponding to edge (i, j).

Basically, that's it and the bipartite matching problem is easily modeled as the following linear program.

$$
\begin{aligned}
\min \sum_{\substack{i=1 \\ j=1}}^{n} & c_{ij} x_{ij} \\
\forall i = 1, \ldots n \ \sum_{j=1}^{n} x_{ij} &= 1 \\
\forall j = 1, \ldots n \ \sum_{i=1}^{n} x_{ij} &= 1 \\
x_{ij} &\geq 0
\end{aligned}
\tag{9.1}
$$

The reader should convince himself that any solution of this program that is integer must be the incidence vector of a perfect matching. What makes life easier compared to the non-bipartite case is that the polyhedron described by the feasible region of this program already is the convex hull of the incidence vectors of perfect matchings as we will show now:

Theorem 34 *Let*

$$P := \{x \in \mathbb{R}^{n^2} \mid x \text{ is feasible for } (9.1)\}.$$

Then $x \in P$ if, and only if, there exist perfect matchings $M_i = (V, F_i)$ and coefficients $\lambda_i \in [0, 1]$, $i = 1, \ldots, k$, such that $\sum_{i=1}^{k} \lambda_i = 1$ and $x = \sum_{i=1}^{k} \lambda_i \chi(F_i)$. Thus P is the convex hull of the incidence vectors of perfect matchings.

Proof   Obviously, all incidence vectors of perfect matchings are feasible for the program and, thus, the same holds for their convex hull by Lemma 7.

Assume on the other hand that x is feasible. We proceed by induction on the size of the support of x.

Consider the *support graph* $H = (V, X)$ of x, where $(i, j) \in X \Leftrightarrow x_{ij} \neq 0$. We claim that it has a perfect matching. Otherwise, by König's Theorem 28 there would exist a vertex cover C such that $|C| < n$. As C is a vertex cover, it contains at least one of the two vertices of each edge, and thus in the following sum every edge appears at least once in the summation on the left and at most once in the summation on the right side:

$$\sum_{\substack{v \in C \cap V_1}} \sum_{\substack{w \in V_2 \\ vw \in X}} x_{vw} + \sum_{\substack{w \in C \cap V_2}} \sum_{\substack{v \in V_1 \\ vw \in X}} x_{vw} \geq \sum_{\substack{v \in V_1 \setminus C}} \sum_{\substack{w \in V_2 \\ vw \in X}} x_{vw} + \sum_{\substack{w \in V_2 \setminus C}} \sum_{\substack{v \in V_1 \\ vw \in X}} x_{vw}.$$

Taking into account that $|V| = 2n$ and using (9.1) we compute

$$n > |C| = \sum_{v \in C} \sum_{w \in V} x_{vw} \geq \sum_{v \in V \setminus C} \sum_{w \in V} x_{vw} = |V| \setminus |C| > n, \tag{9.2}$$

a contradiction. Thus, H contains a perfect matching M.

Let

$$\alpha := \min_{(ij) \in M} x_{ij} > 0.$$

If $\alpha = 1$, then $x = \chi(M)$, and we are done. Otherwise, let

$$\widetilde{x} := \frac{1}{1 - \alpha}(x - \alpha \chi(M)) \quad \geq 0.$$

Then, for all i

$$\sum_{j=1}^{n} \widetilde{x}_{ij} = \frac{1}{1 - \alpha} \sum_{j=1}^{n}(x_{ij} - \alpha \chi(M)_{ij}))$$

$$= \frac{1}{1 - \alpha}(1 - \alpha \cdot 1) = 1,$$

and also $\sum_{i=1}^{n} \widetilde{x}_{ij} = 1$ for all j. Hence $\widetilde{x} \in P$, and $\widetilde{x}$ has a smaller support. Therefore, by induction there exist perfect matchings $M_1 \ldots, M_l$ and coefficients $\lambda_1, \ldots, \lambda_l \geq 0$ such that $\widetilde{x} = \sum_{i=1}^{l} \lambda_i \chi(M_i)$ and $\sum_{i=1}^{l} \lambda_i = 1$. Thus verifying

$$x = \alpha \chi(M) + \sum_{i=1}^{l} \lambda_i (1 - \alpha) \chi(M_i)$$

we find that x is a convex combination of characteristic vectors of perfect matchings.

□

If there is no fractional edge, then x itself is the incidence vector of a matching, so assume $0 < x_e < 1$. Obviously, any vertex incident to a fractional edge must be incident with at least a second fractional edge:

Exercise 69 Give an alternative proof of Theorem 34, by showing that all vertices of P are integral. Given a fractional point x from P find a circuit in its support graph and use it to write x as proper convex combination of two points from P.

As we have turned the weighted bipartite matching into a linear program, we may apply duality. In the case of maximum weighted bipartite matching this yields a "weighted version" of König's Theorem. In case of minimization, the inequality in the non-trivial restrictions is ≤ instead of ≥, we say it is a *packing* instead of a *covering* constraint.

Let us discuss how we derive the dual program in detail. By Exercise 27 the equality constraints correspond to unrestricted variables in the dual program. Apart from that (9.1) already has the shape of (4.6).

Thus, our dual program becomes:

$$
\begin{aligned}
\max \ & \sum_{i=1}^{n} u_i + \sum_{j=1}^{n} v_j \\
\text{s.t.} \quad & u_i + v_j \le c_{ij} \qquad \forall (i, j) \in E.
\end{aligned}
\tag{9.3}
$$

This dual program has a nice interpretation in the case of geometric bipartite matching. Imagine a graph consisting of red and blue points in the plane, and weights c_{ij} that are given by the Euclidean distances between points i and j of different color (see Fig. 9.2 left). The dual variables are vertex weights. We may visualize them as red and blue bubbles or moats around the points (see Fig. 9.2 right). The condition $\forall (i, j) : u_i + v_j \le c_{ij}$ translates to requiring that a red and a blue bubble may touch but never intersect. Our objective is to maximize the sum of the radii of the bubbles.

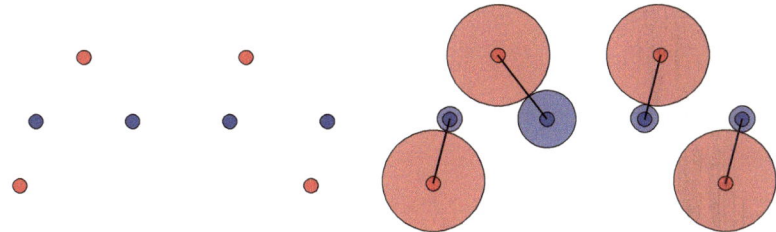

Fig. 9.2 Geometric bipartite matching duality

Let us examine weak duality here. Let u, v be feasible for the dual and x feasible for the primal program. Then using (9.1) we compute

$$\sum_{i=1}^{n} u_i + \sum_{j=1}^{n} v_j = \sum_{\substack{i=1 \\ j=1}}^{n} u_i x_{ij} + \sum_{\substack{i=1 \\ j=1}}^{n} v_j x_{ij}$$

$$= \sum_{\substack{i=1 \\ j=1}}^{n} (u_i + v_j) x_{ij}$$

$$\leq \sum_{\substack{i=1 \\ j=1}}^{n} c_{ij} x_{ij}.$$

Recall that complementary slackness holds for any pair of primal and dual optimal solutions. In the primal dual approach we search for a solution that is complementary to the present dual solution. As long as this primal solution is infeasible we modify the dual solution accordingly. If u and v are feasible radii for the dual program, we call a primal x complementary if it has a non-zero entry only if $u_i + v_j \leq c_{ij}$ is satisfied with equality, a matching is complementary if it consists solely of edges where a red and a blue bubble touch, in other words it only contains tight edges. Thus, we consider the bipartite graph of these tight edges: $G = (V_1 \dot\cup V_2, D)$ where $D = \{(i, j) \mid u_i + v_j = c_{ij}\}$. Obviously, the incidence vector of any matching of this graph is complementary to u and v. If G has a perfect matching, this incidence vector belongs to a feasible, complementary solution, and we have found a dual pair of optimal solutions.

Otherwise, if G has no perfect matching, König's Theorem guarantees the existence of a vertex cover C with $|C| < n$, and our labeling-algorithm 8.1 will find one. As C is a cover, no red bubble around a vertex from $V \setminus C$ touches a blue bubble around a vertex from $V \setminus C$. Since $|C| < n < |V| \setminus C$, enlarging the bubbles around $V \setminus C$ and decreasing the radii of those around C yields a feasible solution with a larger objective value. More precisely, we set $\alpha = \min\{c_{ij} - u_i - v_j \mid i, j \notin C\} > 0$ and choose $\varepsilon \in [0, \alpha]$, and

$$u_i^\varepsilon := \begin{cases} u_i + \varepsilon & \text{if } i \notin C \\ u_i - (\alpha - \varepsilon) & \text{if } i \in C \end{cases} \qquad v_j^\varepsilon := \begin{cases} v_j + (\alpha - \varepsilon) & \text{if } j \notin C \\ v_j - \varepsilon & \text{if } j \in C. \end{cases}$$

We check that $(u^\varepsilon, v^\varepsilon)$ is feasible for the dual program:

$$u_i^\varepsilon + v_j^\varepsilon = \begin{cases} u_i + \varepsilon + v_j + (\alpha - \varepsilon) = u_i + v_j + \alpha \leq c_{ij} & \text{if } i \notin C, \ j \notin C \\ u_i + \varepsilon + v_j - \varepsilon = u_i + v_j \leq c_{ij} & \text{if } i \notin C, \ j \in C \\ u_i - (\alpha - \varepsilon) + v_j + (\alpha - \varepsilon) = u_i + v_j \leq c_{ij} & \text{if } i \in C, \ j \notin C \\ u_i - (\alpha - \varepsilon) + v_j - \varepsilon = u_i + v_j - \alpha < c_{ij} & \text{if } i \in C, \ j \in C. \end{cases}$$

$$(9.4)$$

On the other hand we have increased the value of the objective function, as

$$\sum_{i=1}^{n} u_i^{\varepsilon} + \sum_{j=1}^{n} v_j^{\varepsilon} = \sum_{\substack{i=1 \\ u_i \notin C}}^{n}(u_i + \varepsilon) + \sum_{\substack{i=1 \\ u_i \in C}}^{n}(u_i - \alpha + \varepsilon) + \sum_{\substack{j=1 \\ v_j \in C}}^{n}(v_j - \varepsilon)$$

$$+ \sum_{\substack{j=1 \\ v_j \notin C}}^{n}(v_j + \alpha - \varepsilon)$$

$$= \sum_{i=1}^{n} u_i + \sum_{j=1}^{n} v_j - |C \cap V_1|\alpha + |V_2 \setminus C|\alpha$$

$$= \sum_{i=1}^{n} u_i + \sum_{j=1}^{n} v_j + (n - |C|)\alpha$$

$$> \sum_{i=1}^{n} u_i + \sum_{j=1}^{n} v_j.$$

Natural choices for ε are $\varepsilon \in \{0, \alpha, \frac{1}{2}\alpha\}$. A choice of $\varepsilon = 0$ or $\varepsilon = \alpha$ yields the *Hungarian Method*. For sake of symmetry here we have chosen the third alternative.

As, by assumption, $c \geq 0$ in the beginning we may start with the feasible solution $u = v = 0$. Altogether, with a choice of $\varepsilon = \frac{1}{2}\alpha$, this yields the following sketch of an algorithm:

ALGORITHM Weighted Bipartite Matching

```
G = (V₁∪̇V₂, D) where D = {(i, j) | uᵢ + vⱼ = cᵢⱼ}
(M, C) = MaximalMatchingMinCover (G)
while |M| ≠ n :
    ε = ½ min{cᵢⱼ − uᵢ − vⱼ | i, j ∉ C} > 0
    for all i ∈ V₁:
        if i ∈ C:
            uᵢ = uᵢ − ε
        else:
            uᵢ = uᵢ + ε
    for all j ∈ V₂:
        if j ∈ C:
            vⱼ := vⱼ − ε
        else:
            vⱼ := vⱼ + ε
    G = (V₁∪̇V₂, D) mit D = {(i, j) | uᵢ + vⱼ = cᵢⱼ}
    (M,C) = MaximalMatchingMinCover (G)
```

In summary, our strategy is as follows: we compute a matching of maximal cardinality in the graph of tight edges. If it is not perfect, we use the minimum vertex cover to update the dual variables. By construction the dual variables (u, v) remain feasible throughout the algorithm.

The analysis of the running time is not too difficult. Consider the situation after an iteration of the outer loop. The vertex cover C that our algorithm constructs consists of the unlabeled vertices in V_1 and the labeled vertices in V_2. In the augmentation by (9.4) we will never loose an edge connecting a labeled vertex from V_1 to a labeled V_2 vertex. Furthermore, all labeled vertices in V_2 are matched. New edges arising after a change of the dual variables connect a labeled vertex from V_1 to an unlabeled vertex from V_2. Therefore, we may resume our old labeling until an augmentation step occurs. We need at most $|V|$ dual updates until an augmentation step occurs and thus the method is polynomial and we can implement it in a straightforward fashion in $O(|V|^2|E|)$. This can be improved to $O(|V|^3)$ by using methods similar to those discussed in Sect. 3.5.

Just like the way we proceeded in the unweighted case, where we modeled the bipartite matching problem as a max-flow problem, we can turn the weighted bipartite matching problem into a mincost flow problem.

Again we orient all edges from V_1 to V_2. All vertices from V_1 supply a unit and all vertices from V_2 have a demand of one. This yields the following linear program,

$$\min \sum_{\substack{i=1 \\ j=1}}^{n} c_{ij} x_{ij}$$

$$\forall i = 1, \ldots, n : -\sum_{j=1}^{n} x_{ij} = -1 \tag{9.5}$$

$$\forall j = 1, \ldots, n : \sum_{i=1}^{n} x_{ij} = 1$$

$$x_{ij} \geq 0.$$

It is possible to apply specialized versions of the algorithms discussed in Chap. 7 to this particular problem. Note that the number of augmentation steps in a successive shortest path algorithm applied to a weighted bipartite matching problem on K_{nn} is exactly n. This yields

Remark 12 Let $S(|V|, |E|)$ denote the time to solve a single source shortest path problem. Then it is possible to compute a minimal weighted bipartite matching in $O(|V|S(|V|, |E|))$ (see [10] for a reference).

Software Exercise 70 Apply the algorithm `SuccessiveShortestPath.alg` from the directory `07-MinCostFlows` to the graph `bipmatch.cat` in `09-WeightedMatching`. This basically models the instance from Fig. 9.2. The cost of the arcs is the Euclidean distance and all arcs have capacity 1.

The numbering of the vertices causes this algorithm first to match vertex 6 to vertex number 8. After a dual update, edge $(6, 7)$ has negative reduced cost. Thus, the interpretation of Successive-Shortest-Paths in terms of Weighted Bipartite Matching is that it is always complementary but neither necessarily primally nor dually feasible. In the end, we find the expected minimum weight matching.

If we add a unique pair of a source and a sink, the instance from Fig. 9.2 turns into `bipmatch2.cat`. The source s now is the only supply vertex and all arcs emanating from it have unit capacity and zero cost. The same holds for the arcs that end in the unique sink t. Since we always consider shortest paths from s to t only, the potentials model the distance from s, and thus the reduced cost of the arcs will stay non-negative, i.e. we stay dually feasible.

9.1.1 Inverse Minimum Spanning Tree Problem

As an application of weighted bipartite matching we discuss the inverse minimum spanning tree problem [45].

Given a weighted graph $G = (V, E, c)$ and a tree T spanning G, the objective is to modify the cost function as carefully as possible such that T becomes a minimal spanning tree. Clearly, for that purpose we will never increase the weight of an edge of T nor decrease the cost of an edge outside of T. Thus we can express the modified function using a non-negative perturbation d as

$$\sum_{e \in T} (c(e) - d(e)) + \sum_{e \notin T} (c(e) + d(e)). \tag{9.6}$$

By the circuit criterion from Lemma 3, T is optimal with respect to the modified cost function if, and only if,

$$\forall f \notin T \forall e \in C(f, T) \setminus \{f\} : c(f) + d(f) \geq c(e) - d(e), \tag{9.7}$$

where $C(f, T)$ is the fundamental circuit that f closes with T, or equivalently

$$\forall f \notin T \ \forall e \in C(f, T) : d(e) + d(f) \geq c(e) - c(f). \tag{9.8}$$

This might remind you of the linear program in (9.3). We want to minimize $\sum_{e \in E} d(e)$ such that a weight, namely $c(e) - c(f)$ is covered. Thus, we consider the bipartite graph defined on the vertex set $T \dot\cup (E \setminus T)$ where (e, f) is an edge of weight $c(e) - c(f)$ if and only if $e \in C(T, f)$. We can solve this problem using the weighted bipartite matching algorithm.

9.2 Non-Bipartite Matching

Finally, we turn to the weighted matching problems on general graphs. As argued in 9.1 we can restrict our considerations to complete graphs K_n with n even and with a non-negative weight function.

Nevertheless the problem is a lot more complicated, as the convex hull of incidence vectors of the matchings does not have such a nice description anymore. Consider for example the graph in Fig. 9.3.

If we consider the analogous program[1] to (9.1),

$$\min \sum_{1 \leq i < j \leq n} c_{ij} x_{ij}$$
$$\forall i = 1, \ldots n \quad \sum_{\substack{j=1 \\ i \neq j}}^{n} x_{ij} \qquad = 1 \qquad\qquad (9.9)$$
$$x_{ij} \qquad\qquad \geq 0,$$

and require the solution to be integer, we will get matchings only. Unfortunately, there are some fractional solutions that are feasible for the above program which are not in the convex hull of the incidence vectors of matchings.

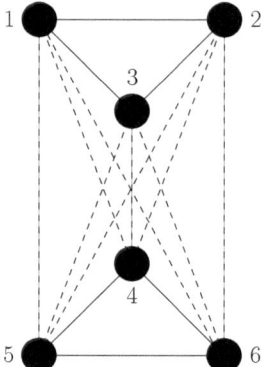

Fig. 9.3 Odd circuits

The graph in Fig. 9.3 depicts the K_6. The vector $x = x_{ij}$ defined by $x_{12} = x_{23} = x_{31} = \frac{1}{2} = x_{45} = x_{56} = x_{64}$ and $x_{ij} = 0$ else is a feasible solution to (9.9). It cannot be a linear combination of incidence vectors of matchings though. Any matching must match at least one node of the upper triangle to one node of the lower triangle 123. Therefore, any convex combination of matchings must have a non-zero entry for some edge in the cut induced by one of these triangles. This proves that the polyhedron described by the above inequalities has vertices with coordinates that are not integer.

9.2.1 The Matching Polytope

We must somehow get rid of fractional vertices, in particular of solutions corresponding to disjoint unions of matching edges and half-integral odd circuits. We

---

[1] To simplify notation we use x_{ij} as well as x_{ji} for the one variable that represents the edge connecting i and j.

achieve this by adding what we call "blossom constraints". We require that the sum of variable entries of any cut defined by an odd set of vertices is at least 1. Recall that $\partial(S)$ denotes the set of edges that have exactly one endpoint in S and consider the following program:

$$
\begin{aligned}
\min \ & \textstyle\sum_{i=1,i<j}^{n} c_{ij} x_{ij} \\
\forall v \in V \ & \textstyle\sum_{w \neq v} x_{vw} && = 1 \\
\forall S \subseteq V, |S| \text{ odd} \ & \textstyle\sum_{e \in \partial(S)} x_e && \geq 1 \\
& x_{ij} && \geq 0.
\end{aligned}
\tag{9.10}
$$

Here and in the following when we consider sets where $|S|$ is odd, we implicitly will assume that $3 \leq |S| \leq |V| - 3$ to distinguish the blossom constraints from the equations for the vertices. In the following theorem we show that the blossom constraints guarantee integrality of the vertices.

Theorem 35 *Let*

$$
P := \{x \in \mathbb{R}^{n^2} \mid x \text{ is feasible for } (9.10)\}.
$$

Then $x \in P$ if, and only if, there exist perfect matchings $M_i = (V, F_i)$ and coefficients $\kappa_i \in [0, 1]$, $i = 1, \ldots, k$, such that $\sum_{i=1}^{k} \kappa_i = 1$ and $x = \sum_{i=1}^{k} \kappa_i \chi(F_i)$. Hence P is the convex hull of the incidence vectors of perfect matchings.

Instead of $\chi(F_i)$ we will also simply write $\chi(M_i)$.

Proof We proceed as in the proof of Theorem 3. One inclusion is obvious, all incidence vectors of perfect matchings are in P, and thus is their convex hull.

Now let $x \in P$. We proceed by induction on $|\operatorname{supp}(x)|$. There is noting to prove, if $|\operatorname{supp}(x)| = \frac{|V|}{2}$. In the following we assume $|\operatorname{supp}(x)| > \frac{|V|}{2}$.

We distinguish two cases. In the first case we have a tight blossom constraint for some $S \subset V$. Then we can contract the blossom and apply induction. Furthermore, we can contract the complement of S, apply induction and combine the convex combinations found in these two steps properly. This will be more or less straightforward but quite tedious.

In the second case none of the blossom constraints is tight. Then $\operatorname{supp}(x)$ contains a perfect matching and, similarly to the previous polyhedral results, we subtract some fraction of this matching from x and apply induction.

Case 1: First we consider the case that there is an odd set $S \subseteq V$ of vertices – recall that this in particular means $3 \leq |S| \leq |V| - 3$ – such that

$$
\sum_{e \in \partial(S)} x_e = 1.
\tag{9.11}
$$

Let H_1 denote the set of edges that have both end points in $V \setminus S$ and H_2 those with both end points in S. We consider the graphs $G_1 := G/H_1$ and $G_2 := G/H_2$. Let x^1 denote the vector derived from x by restricting it to $H_2 \cup \partial(S)$ and possibly summing values of edges of G that are parallel in G_1 into a single value of an edge of G_1. Then $x^1 \in \mathbb{R}^{E(G_1)}$.

We claim that x^1 is feasible for the matching program (9.10) for G_1. This is obvious for the equalities for original vertices as well as for the blossom constraints. For the new super node s in G_1 it holds by (9.11). Since

$$2 \sum_{e \in H_1} x_e = \sum_{i,j \in V \setminus S} x_{ij} = \sum_{i \in V \setminus S} \sum_{\substack{j=1 \\ j \neq i}}^{n} x_{ij} - \sum_{e \in \partial(S)} x_e = |V \setminus S| - 1 \geq 2 > 0,$$

our x^1 has smaller support than x. Thus, we can apply the inductive assumption to conclude that x^1 is a convex combination

$$x^1 = \sum_{i=1}^{k} \lambda_i \chi_{\tilde{N}_i}$$

of incidence vectors of perfect matchings $\tilde{N}_1, \dots, \tilde{N}_k$ of G_1. The same holds for x^2, the restriction of x to $H_1 \cup \partial(S)$, and perfect matchings $\tilde{O}_1, \dots, \tilde{O}_l$ of G_2:

$$x^2 = \sum_{j=1}^{l} \mu_j \chi_{\tilde{O}_j}.$$

As $\tilde{N}_i$ is a perfect matching the super node s has exactly one matching partner u_i in S. Omitting (su_i) yields a matching N_i that matches all but one vertex, namely u_i, from S. Similarly, O_j, derived from $\tilde{O}_j$, matches all but one vertex, namely v_j of $V \setminus S$. We can combine N_i and O_j into a matching M_{ij} of G by matching u_i to v_j, using N_i inside of S and O_j inside of $V \setminus S$. In the next step we construct the coefficients for the $\chi_{M_{ij}}$.

Recall that, given $i \in \{1, \dots, k\}$, $u_i \in S$ denotes the unique vertex that is unmatched in N_i, and $v_j \in V \setminus S$ is the unique unmatched vertex in O_j. Then $u_i v_j$ is the only edge in $M_{ij} \cap \partial(S)$. Note that the same vertex can be unmatched in different N_is or O_js. That is we may have $u_{i_1} = u_{i_2}$ for $i_1 \neq i_2$ respectively $v_{j_1} = v_{j_2}$ for $j_1 \neq j_2$.

For fixed $u \in S$ and $u_i = u$ we may combine N_i with all matchings O_j. How to derive the coefficient for the resulting perfect matching $N_i \cup O_j \cup (uv)$ for $v = v_j$ (see Fig. 9.4)? We have to ensure that the final convex combination yields the proper weight on the edges $(uv) \in \partial S$. Considering all matchings N_i that leave u unmatched, we conclude that

$$\sum_{u \in e \in \partial(S)} x_e = \sum_{u_o = u} \lambda_o \tag{9.12}$$

and hence the fraction of $\lambda_i \chi_{N_i}$ that we should combine with some O_j that leaves v unmatched should be

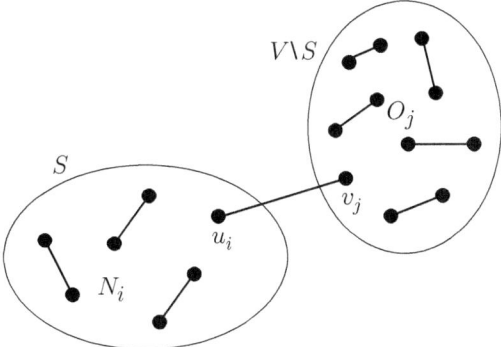

Fig. 9.4 Combining two matchings

$$\frac{\lambda_i}{\left(\sum_{u_o=u} \lambda_o\right)} x_{uv}.$$

Note that the sum is taken over all vertices u_o satisfying $u_o = u$. Similar considerations hold for $v = v_j$ and $\mu_j \chi_{O_j}$. Thus we define

$$\kappa_{ij} = \frac{\lambda_i}{\left(\sum_{u_o=u} \lambda_o\right)} \cdot \frac{\mu_j}{\left(\sum_{v_p=v} \mu_p\right)} x_{uv}$$

and claim that

$$x = \sum_{i=1}^{k} \sum_{j=1}^{l} \kappa_{ij} \chi_{M_{ij}}$$

is a convex combination of characteristic vectors of perfect matchings. Recall that κ_{ij} is the coefficient of the perfect matching $N_i \cup O_j \cup (u_i v_j)$.

Clearly, the κ_{ij} are non-negative and

$$\sum_{i=1}^{k} \sum_{j=1}^{l} \kappa_{ij} = \sum_{u \in S} \sum_{v \in V \setminus S} \left(\sum_{u_i=u} \sum_{v_j=v} \frac{\lambda_i \mu_j x_{uv}}{\left(\sum_{u_o=u} \lambda_o\right) \left(\sum_{v_p=v} \mu_p\right)} \right)$$

$$= \sum_{u \in S} \sum_{v \in V \setminus S} x_{uv}$$

$$= \sum_{(uv) \in \partial(S)} x_{uv} = 1.$$

In order to verify that $x_e = \sum_{i=1}^{k} \sum_{j=1}^{l} \kappa_{ij} \chi_{M_{ij}}(e)$ for all $e = (u'v') \in E$ we distinguish the cases $(u'v') \in \partial(S)$, $\{u', v'\} \subseteq S$ and $\{u', v'\} \subseteq V \setminus S$. The last two cases are symmetric, thus we may restrict our attention to the first two.

First, let $e = (u'v') \in \partial(S)$. Then $\chi_{M_{ij}}(e) = 1$ iff $u' = u_i$ and $v' = v_j$, and zero else, and thus

$$\sum_{i=1}^{k}\sum_{j=1}^{l} \kappa_{ij}\chi_{M_{ij}}(e) = \sum_{u'=u_i}\sum_{v'=v_j} \kappa_{ij}$$

$$= \sum_{u'=u_i}\sum_{v'=v_j} \frac{\lambda_i \mu_j x_e}{\left(\sum_{u_o=u}\lambda_o\right)\left(\sum_{v_p=v}\mu_p\right)} = x_e.$$

Second, let $e = (u'v')$ and $u', v' \in S$. Then for all i, j we have $\chi_{M_{ij}}(e) = \chi_{N_i}(e)$ and furthermore, since a vertex $u \in S$ is matched internally unless $u_i = u$, as in (9.12) we necessarily must have

$$\sum_{u_i=u'} \lambda_i = \sum_{u' \in e \in \partial(S)} x_e = \sum_{v \in V \setminus S} x_{u'v}.$$

Using that we compute

$$\sum_{i=1}^{k}\sum_{j=1}^{l} \kappa_{ij}\chi_{M_{ij}}(e) = \sum_{u \in S}\sum_{v \in V \setminus S}\sum_{u_i=u}\sum_{v_j=v} \frac{\lambda_i \mu_j x_{uv}}{\left(\sum_{u_o=u}\lambda_o\right)\left(\sum_{v_p=v}\mu_p\right)}\chi_{N_i}(e)$$

$$= \sum_{u \in S}\sum_{v \in V \setminus S}\sum_{u_i=u} \frac{\lambda_i \left(\sum_{v_j=v}\mu_j\right) x_{uv}}{\left(\sum_{u_o=u}\lambda_o\right)\left(\sum_{v_p=v}\mu_p\right)}\chi_{N_i}(e)$$

$$= \sum_{u \in S}\sum_{v \in V \setminus S} \frac{x_{uv}}{\left(\sum_{u_o=u}\lambda_o\right)}\sum_{u_i=u}\lambda_i \chi_{N_i}(e)$$

$$= \sum_{u \in S} \frac{\left(\sum_{v \in V \setminus S}x_{uv}\right)}{\left(\sum_{u_o=u}\lambda_o\right)}\sum_{u_i=u}\lambda_i \chi_{N_i}(e)$$

$$= \sum_{u \in S}\sum_{u_i=u}\lambda_i \chi_{N_i}(e)$$

$$= \sum_{i=1}^{k}\lambda_i \chi_{N_i}(e)$$

$$= x_e^1 = x_e.$$

Thus we are done with Case 1.

Case 2: For all odd sets $S \subseteq V$ of vertices satisfying $3 \leq |S| \leq |V| - 3$ we have

$$\sum_{e \in \partial(S)} x_e > 1.$$

We show that the support graph H of x contains a perfect matching by contradiction. Assume that H contains no perfect matching, then by Theorem 33 there exists a barrier A such that $|A| < k$ and $H \setminus A$ has k odd components $H_1, \ldots, H_k$. Due to the blossom constraints we know that

$$\sum_{i=1}^{k} \sum_{e \in \partial(H_i)} x_e \geq k. \tag{9.13}$$

On the other hand

$$\sum_{i=1}^{k} \sum_{e \in \partial(H_i)} x_e \leq \sum_{e \in \partial(A)} x_e \leq |A| < k, \tag{9.14}$$

a contradiction. Thus, H contains a perfect matching M_0.

Now, we proceed as in the proof of Theorem 34 and carefully subtract a multiple of this matching until either a variable drops to zero or a blossom constraint becomes tight. To be precise let

$$\kappa_0 = \min \left\{ \{x_e \mid e \in M_0\}, \max\{\varepsilon \mid \forall |S| \text{ odd } \sum_{e \in \partial(S)} (x_e - \varepsilon \chi_{M_0}(e)) \geq (1 - \varepsilon)\} \right\}.$$

If $\kappa_0 = 1$, then x is the incidence vector of a perfect matching and we are done. Otherwise, similar to the analogous part in the proof of Theorem 3, we set

$$\tilde{x} := \frac{1}{1 - \kappa_0} (x - \kappa_0 \chi_{M_0}).$$

By construction $\tilde{x}$ is non-negative. Let $S \subseteq V$ be of odd cardinality. Then

$$\sum_{e \in \partial(S)} (x_e - \kappa_0 \chi_{M_0}(e)) \geq (1 - \kappa_0) \Rightarrow \sum_{e \in \partial(S)} \tilde{x}_e \geq 1.$$

Hence, it satisfies the blossom constraints. If $u \in V$ then

$$\sum_{v \in V} \tilde{x}_{uv} = \frac{1}{1 - \kappa_0} \left(\sum_{v \in V} x_{uv} - \kappa_0 \right) = \frac{1 - \kappa_0}{1 - \kappa_0} = 1.$$

Thus $\tilde{x}$ is feasible for (9.10). Moreover, either it has a smaller support than x or there is some odd set S, $3 \leq |S| \leq |V| - 3$ such that

$$\sum_{e \in \partial(S)} (x_e - \kappa_0 \chi_{M_0}(e)) = (1 - \kappa_0) \iff \sum_{e \in \partial(S)} \tilde{x}_e = 1.$$

Thus, either by Case 1 or by inductive assumption there are perfect matchings $M_1, \ldots, M_k$ and convex coefficients $\widetilde{\kappa}_1, \ldots, \widetilde{\kappa}_k$ such that

$$\widetilde{x} = \sum_{i=1}^{k} \widetilde{\kappa}_i \chi_{M_i}.$$

Now we set $\kappa_i = (1 - \kappa_0)\widetilde{\kappa}_i$ to get

$$\sum_{i=0}^{k} \kappa_i \chi_{M_i} = \kappa_0 \chi_{M_0} + (1 - \kappa_0) \sum_{i=1}^{k} \widetilde{\kappa}_i \chi_{M_i}$$
$$= \kappa_0 \chi_{M_0} + (1 - \kappa_0)\widetilde{x} = x.$$

$\square$

9.2.2 The Weighted Matching Algorithm

Thus, (9.10) properly models the weighted matching problem as a linear program. In order to apply our machinery of primal-dual algorithms we first consider the corresponding dual program. We have a non-negative variable U_S for each blossom constraint and variables u_i without sign restriction for the vertices. Altogether the dual program reads:

$$\begin{aligned} \max \quad & \sum_{i=1}^{n} u_i + \sum_{\substack{S \subseteq V \\ S \text{ odd}}} U_S \\ \forall (i, j) \ & u_i + u_j + \sum_{(i,j) \in \partial(S)} U_S \leq c_{ij} \\ & U_S \geq 0. \end{aligned} \qquad (9.15)$$

Following our approach to construct pairs of a dually feasible and a—not necessarily feasible—complementary primal solution we consider a dually feasible solution (u, v) and compute a maximum matching in the graph D of the tight edges. If this matching is perfect, we are done. Otherwise, by Theorem 33 we find a barrier A and odd components $H_1, \ldots, H_k$ of $D \setminus A$ such that $k > |A|$. As we have variables U_{H_i} for the odd components, increasing their U_{H_i} while decreasing the variables corresponding to the nodes or super nodes in A will improve our dual objective function. We can do so until some other constraint becomes strict.

Let us for a moment return to the discussion of the blossom algorithm for the unweighted case. There, after each augmentation, we had restarted the algorithm with all blossoms expanded. This way we could easily derive the Tutte-Berge formula. Here, we have to be a bit more careful. Note, that there are exponentially many dual variables. In order to derive a polynomial time algorithm we should use only a polynomial number of them. This is guaranteed, if the set family corresponding to the non-zero variables is *nested*.

Definition 32 Let $\mathcal{F} \subseteq 2^E$ be a set family of non-empty sets. We say the family is *nested* if for all $S, S' \in \mathcal{F}$ holds: $S \cap S' \neq \emptyset \Rightarrow (S \subset S'$ or $S' \subset S)$.

Proposition 9 *Let $\mathcal{F} \subseteq 2^E$ be a nested family. Then $|\mathcal{F}| \leq 2|E| - 1$.*

Proof   We proceed by induction on $|E|$. If $|E| = 1$, there is nothing to prove, thus let $|E| \geq 2$ and $e \in E$. We consider the set family $\mathcal{F} \setminus e := \{S \setminus e \mid S \in \mathcal{F}\}$, removing duplicates and, possibly, the empty set. Obviously, this is a nested family and by induction we conclude that it consists of at most $2|E| - 3$ sets. Now assume that T_1, T_2 are non-empty sets in $\mathcal{F}$ not containing e such that $T_1 \cup \{e\}, T_2 \cup \{e\} \in \mathcal{F}$. Since $\mathcal{F}$ is nested, we may assume $T_1 \subset T_2$. But then $(T_1 \cup e) \cap T_2 = T_1 \neq \emptyset$ and neither $T_1 \cup \{e\} \subseteq T_2$ nor $T_2 \subseteq T_1 \cup \{e\}$, contradicting $\mathcal{F}$ being nested. Therefore, there can be at most one such non-empty set. Furthermore, we may have $\{e\} \in \mathcal{F}$ and therefore $|\mathcal{F}| \leq 2|E| - 3 + 2 = 2|E| - 1$. $\square$

The following proposition will guarantee that it is possible to restrict the non-zero dual variables to a nested family. It shows that in a certain sense we can contract blossoms similarly to the way we proceeded in the unweighted case.

Proposition 10 *Let (u, v) be dually feasible for a matching problem (K_n, c) and C be an odd circuit of tight edges. Let S_C denote the set of vertices of C. We define a weight function on K_n / C as follows:*

$$c'_e := \begin{cases} c_e & \text{if } e \text{ is not incident with a vertex of } C \\ c_e - u_{v_1} & \text{if } \quad e \in \partial(V(C)) \text{ and } v_1 \in V(C). \end{cases}$$

Let M' be an optimal matching for the problem $(K_n / C, c')$ and (u', v') be feasible for the corresponding dual program, such that $u'_C \geq 0$ and M' consists of tight edges only. We complete M' to a perfect matching M of G using edges from C and put

$$\tilde{u} := \begin{cases} u_v & \text{if } v \in V(C) \\ u'_v & \text{if } v \notin V(C) \end{cases} \qquad \tilde{U}_S := \begin{cases} U'_S & \text{if } U'_S > 0 \text{ and } S \neq S_C \\ u'_{S_C} & \text{if } \qquad\qquad S = S_C. \end{cases}$$

Then $(\tilde{u}, \tilde{U})$ is feasible (9.15) and M is a perfect matching of the original problem using only edges tight with respect to $\tilde{u}$.

Proof   First we check feasibility of $(\tilde{u}, \tilde{U})$. By assumption $\tilde{U}_{S_C}$ is non-negative, the other non-negativity constraints are guaranteed by feasibility of u'. For the remaining constraints we have to consider the cases $(i, j) \in \partial(S_C)$, $\{i, j\} \subseteq V \setminus S_C$ and $\{i, j\} \subseteq C$. In the last case the constraints are satisfied since u is feasible, and in the second case since u' is feasible. Thus, let $(ij) \in \partial(S_C)$ and w.l.o.g. $i \in S_C$, then

$$\tilde{u}_i + \tilde{u}_j + \sum_{\substack{(i,j)\in\partial(S)}} \tilde{U}_S = u_i + u'_j + u'_{S_C} + \sum_{\substack{(i,j)\in\partial(S)\\S\neq S_C}} U'_S$$

$$= u_i + u'_{S_C} + u'_j + \sum_{\substack{(i,j)\in\partial(S)\\S\neq S_C}} U'_S$$

$$\leq u_i + c'_{ij} = u_i + c_{ij} - u_i = c_{ij}.$$

Since C consists of tight edges with respect to u only, we can match all but one vertex of S_C using edges of C which are complementary with respect to $\tilde{u}$. By assumption M' is complementary to u'. Let e be the unique edge of M' which is incident to the super node S_C and let (ij), $i \in S_C$ be the corresponding original edge in G. Clearly, all other edges of M' use only tight edges of $\tilde{u}$. Since e is tight with respect to c'_{ij}, the following computation finishes the proof

$$\tilde{u}_i + \tilde{u}_j + \sum_{\substack{(i,j)\in\partial(S)}} \tilde{U}_S = u_i + u'_j + u'_{S_C} + \sum_{\substack{(i,j)\in\partial(S)\\S\neq S_C}} U'_S$$

$$= u_i + u'_{S_C} + u'_j + \sum_{\substack{(i,j)\in\partial(S)\\S\neq S_C}} U'_S$$

$$= u_i + c'_{ij} = u_i + c_{ij} - u_i = c_{ij}.$$

$\square$

Thus, we may proceed as follows:

(i) Throughout the algorithm we keep in (u, U) a feasible solution of (9.15) such that the sets with non-zero entry in the corresponding variables form a nested family, in the beginning we may start e.g. with the all zero vector. Actually, we will not have explicit variables for the odd sets and will create them only if they become non-zero.

(ii) We compute a maximal matching on the graph of tight edges. If it is not perfect, we compute a barrier A and sets H_i as in Theorem 33.

(iii) In case they do not already exist we create variables for $H_1, \ldots, H_k$ and determine the maximum ε such that increasing the dual variables on the H_i by ε and decreasing them on the variables corresponding to the nodes and super nodes in S does not violate dual feasibility. We have to check for three events:

   (i) An inequality for an edge between two of the odd sets becomes tight. This happens for $\varepsilon = \varepsilon_1$ and

$$\varepsilon_1 = \frac{1}{2} \min\{c_{ij} - u_i - u_j - \sum_{\substack{(i,j)\in\partial(S)}} u_S \mid i \in H_{k_1}, j \in H_{k_2}, 1 \leq k_1 < k_2\}$$

   (ii) An inequality for an edge between one of the odd sets and a vertex that is neither in the odd sets nor in A becomes tight. This happens for $\varepsilon = \varepsilon_2$

and

$$\varepsilon_2 = \min\{c_{ij} - u_i - u_j - \sum_{(i,j)\in\partial(S)} u_S \mid i \in H_{k_1}, j \in H_{k_2}, 0 = k_1 < k_2\}.$$

(iii) A variable for a super node in A drops to zero. This happens for $\varepsilon = \varepsilon_3$ and

$$\varepsilon_3 = \min\{u_v \mid v \text{ is a super node of } G\}.$$

(iv) Now, we compute

$$\varepsilon = \min\{\varepsilon_1, \varepsilon_2, \varepsilon_3\}$$

and set $u_v = u_v - \varepsilon$ for $v \in A$ and $U_{H_i} = U_{H_i} + \varepsilon$ and iterate.

Summarizing this yields the following algorithm:

ALGORITHM WeightedMatching

```
(M, A, B, C) = GallaiEdmondsDecomposition(G)

while not (A == [] and B == [] and C == []):
    for v in B:
        if is_supervertex(v) and not is_dualvar(v):
            u[v] = 0.0   # adds new dual var

    eps1, tight_edges1 = mindist(B, B)
    eps1 *= 0.5
    eps2, tight_edges2 = mindist(B, C)
    eps3 = min([u[v] for v in A if is_supervertex(v)] + [Infinity])
    i, eps = argmin_min([eps1,eps2,eps3])
    tight_edges = [tight_edges1, tight_edges2, []][i]

    for v in A: # ----- deflate A
        u[v] -= eps

    for v in B: # ----- inflate B
        u[v] += eps

    for v in A:
        if not is_supervertex(v):
            RemoveIncidentUntightEdges(v,A,B,C)

    if i == 2: # eps == eps3
        for v in A:
            if is_supervertex(v) and u[v] == 0.0:
                Unshrink(v)

    for v,w in tight_edges:
        AddTightEdge(v,w)

    (M, A, B, C) = GallaiEdmondsDecomposition(G)
```

We use the procedure to compute the Gallai-Edmonds decomposition from Edmonds' Blossom Algorithm. It returns a barrier A, a set of odd components B and a list of remaining vertices C. In the next step we compute $\varepsilon := \min\{\varepsilon_1, \varepsilon_2, \varepsilon_3\}$ update the dual variables and iterate.

As in the bipartite case we have a nice geometric interpretation for the dual variables, in the case of matching points in the Euclidean plane, which is shown in our implementation. The additional variables corresponding to the blossom constraints give rise to variables around odd groups, we may depict them as moats.

Before we analyze the complexity and discuss implementational details, let us watch the algorithm at work.

Software Exercise 71 Go to the directory `09-WeightedMatching` start Gato and load the algorithm `WeightedMatching.alg` and the graph `two triangles.cat`. Our implementation of the weighted matching algorithm computes a minimum weight matching of a complete graph in the plane where the weights are given by Euclidean distances, thus we ignore edges or non-edges that occur in the first window at the start.

In the first step we determine a maximum matching in the graph of tight edges. The routine `GallaiEdmondsDecomposition(G)` returns a barrier A, a set of odd components B and the set of remaining, necessarily matched, vertices C. Since our dual variables initially are set to zero, in the beginning G is empty. Hence A = $\emptyset$ and B = V. Since B contains no non-trivial component, we need not create any new variable. Since A and C are empty, we have eps = eps1 where eps1 is half the length of the edges $(1, 3)$, $(2, 3)$, $(4, 5)$ and $(5, 6)$, which are all equal. We increase the value of all dual variables for the vertices by eps and the above edges become tight. We add these edges to G and start Edmonds' Blossom Algorithm. It returns A = $\{3, 4\}$—we have chosen to paint A red in the second window, the *shrinking window*—and B = $\{1, 2, 5, 6\}$. We decrease u_3 and u_4 and increase u_1, u_2, u_5 and u_6 until the edges $(1, 2)$ and $(5, 6)$ become tight, i.e. eps = eps1, add these edges to G and our blossom algorithm shrinks both triangles, returns A = $\emptyset$ and B = $\{7 = \{1, 2, 3\}, 8 = \{4, 5, 6\}\}$. Now, we have to create the blossom variables U_7 and U_8, and eps = eps1 is half the distance of the small moat around 3 and the large moat around 6 in the first window. We add the edge $(7, 8)$ and Edmonds' Blossom Algorithm returns a perfect matching. All other sets are empty. In the first window you can see that we had to change the matchings inside of the triangles in order to make vertices 3 and 6 unsaturated. We will discuss how we do this in more detail later.

Software Exercise 72 Next we consider the graph `3223.cat`. In the first five rounds we always add a single edge to our graph of tight edges. The dual variables of their end vertices keep their value in the next iterations. In the sixth iteration $A \neq \emptyset$ for the first time and u_1 is decreased. As a result we have a triangle in the upper left which is shrunken to $11 = \{1, 2, 3\}$ in the next iteration where we gain edge $(10, 8)$. In the next iteration for the first time we find eps=eps2 which makes $(2, 7)$ tight. In the next round we create the new super node $12 = \{11, 6, 7\}$ and hence for the first time we meet nested blossom variables.

When we, finally, arrive at a graph with a perfect matching after adding edge $(2, 4)$ we have to change the matching in super node 13 in order to make 2, 4, 5, 9 an augmenting path.

In order to have an example where eps = eps3 we run our algorithm on var2 zero.cat. After three iterations we form the super node $7 = \{1, 2, 6\}$ and create the variable u_7 which in the next iteration is set to a strictly positive value and 7 is matched in the shrunken graph. In the following iteration we have $A = \{7\}$ and we have to decrease u_7. It drops to zero before another tight edge occurs. In our shrunken graph we have to unshrink 7 and the blossom algorithm returns $A = \{2\}$. The edges $(1, 2)$ and $(1, 6)$ become untight and 2 forms a new super node with 3 and 4 which is matched in the next iteration.

Let us do one more experiment:

Software Exercise 73 Load the graph 11vs13.cat. In the first two iterations two perfectly matched components of size eight are formed. You can see that something happens to these components in Edmonds' blossom algorithm. Set a breakpoint on the line (M, A, B, C) = GallaiEdmondsDecomposition(G) and restart the algorithm. If you watch carefully you will see that at the end of the second iteration Edmonds' blossom algorithm shrinks the triangle $\{1, 2, 3\}, \{5, 6, 7\}$, $\{13, 14, 15\}$ and $\{17, 18, 19\}$. Nevertheless, this has no effect on our primal dual procedure since all these triangles after the second iteration belong to perfectly matched components and thus to the set C. At the end of the fourth iteration, when $(10, 8)$ has occurred 8 becomes a vertex in the barrier and $31 = \{1, 2, 3, 4, 5, 6, 7\}$ an odd component in B. This is augmented by 8 and 10 to become the new super node 32. Observe that up to now all edges leaving 32 end in original vertices of the graph. This changes some iterations later, e.g., when $\{15, 16, 17, 18, 19, 20, 21, 22, 24\}$ for the new super node 36, or when the graph in the shrinking window is reduced to the two super nodes 38 and 39. In order to finally match these two vertices we have to find out which edge in the original graph corresponds to $(38, 39)$. To be able to do that requires some thoughts about the appropriate data structure.

There will be several points of discussion in the analysis of the algorithm. The argument for correctness is the same as before in all primal-dual algorithms. Next, let us consider the number of main iterations of our procedure.

Note, that while computing the labeling, we can continue the old labeling after a dual update until an augmentation step occurs. The only edges that we may loose are edges between vertices in the barrier. Since all vertices in the barrier are (matched) inner vertices and thus receive their label via the matching edge we will never loose an edge used in the labeling procedure.

The fact that all vertices in A are inner and have an even label also implies that this holds for all super nodes that we have to expand because their dual variable drops to zero. Since all super nodes that are newly formed are outer and have an odd label, between two augmentations steps, we will never form and expand the same blossom.

We conclude that in a dual update we either expand a blossom or increase the set of labeled vertices. By the above we have at most $O(|V|)$ blossom expansions and, since we can clearly increase the set of labeled vertices at most n times, we perform at most $O(|V|)$ changes of the dual variables until an augmentation step occurs. Hence, our algorithm requires at most $O(|V|^2)$ main iterations.

What is the complexity of computing ε. In the naive implementation that we have chosen in CATBox, we have for each vertex a list of blossoms with non-zero variable that it belongs to. This way we can compute

$$c_{ij} - u_i - u_j - \sum_{(i,j) \in \partial(S)} u_S$$

for each edge (i, j) in $O(|V|)$. Thus, determining ε requires $O(|E||V|) = O(|V|^3)$ since $|E| = \binom{|V|}{2}$ in the complete graph. This still is dominated by the complexity of our implementation of Edmonds' blossom algorithm which is $O(|V|^2|E|)$.

There is a lot of space for improvements. We have to encode the nested structure of the shrunken graphs anyway. In particular, for each edge in the shrunken graph we have to find an appropriate edge when we unshrink a blossom B. This may be complicated by the fact that the other end vertices of edges in ∂B meanwhile may have disappeared in another blossom. One solution is to have a pointer to an original edge for each edge in the shrunken graph and a pointer to the outermost blossom it lies in for each edge.

With each blossom and each edge in its boundary we can store the present value of $c_{ij} - u_i - u_j - \sum_{(i,j) \in \partial(S)} u_S$. This enables us to compute ε in $O(|E|)$. If we recycle the labeling of Edmonds' blossom algorithm, we can reduce the cost of an augmentation to $O(|V||E|)$ and all in all our algorithm has the same complexity as the cardinality matching algorithm.

Remark 13 In the 80s of the last century several improvements to the above have been proposed. The best currently known bounds are $O(|V|(|E| + |V| \log |V|))$ and $O\left(|E| \log(|V|C)\sqrt{|V|\alpha(|E|, |V|)} \log |V|\right)$ where C is an upper bound on the absolute value of the (integer) weights and α is the inverse of the Ackerman function. See [38] for details and a table with a complexity survey.

9.3 Karyotyping and the Chinese Postman Problem

In the first chapter in Applications 6 and 7 we introduced two applications of weighted matching that are not that immediate. We will discuss them in some detail here.

In Application 6 we wanted to map 46 chromosomes to 23 chromosome classes, two to each of them. The probability of probe i to belong to class j is p_{ij}. These probabilities may be a result from some pattern recognition program and are assumed to be independent of one another. Thus, the probability of a map $m(i)$ is given by

$$m(i) = \prod_{i=1}^{46} p_{im(i)}. \tag{9.16}$$

We want to compute the maximum of that function under the restriction that each class is the image of exactly two probes. For that purpose we compute a minimum weight bipartite perfect matching on the $K_{46,46}$ where the weights are given by

$$c_{i,2j-1} = c_{i,2j} = -\log p_{ij} \geq 0.$$

Any solution of this matching problem will map exactly two probes to each class and maximize the function

$$\sum_{i=1}^{46} \log p_{im(i)} = \log\left(\prod_{i=1}^{46} p_{im(i)}\right).$$

As the logarithm is monotone this is as required.

In Application 7, the Chinese Postman Problem, we are given a weighted graph $G = (V, E, w)$ and search for a tour of minimal weight through the graph that visits all edges. Such a tour is called a Eulerian tour.

Definition 33 Let $G = (V, E)$ be a graph. A cycle that visits each edge exactly once is called a *Eulerian tour*. A graph that admits a Eulerian tour is called *Eulerian*.

We have an immediate characterization of Eulerian graphs:

Theorem 36 *Let $G = (V, E)$ be a graph. Then the following is pairwise equivalent.*

 (i) *G is Eulerian.*
 (ii) *G is connected and all vertices have even degree.*
(iii) *G is connected and E is a disjoint union of circuits.*

Proof The first implication is obvious, as for each vertex we need as many edges to enter it as we need to leave it.

The second is shown by induction on $|E|$. Let $G = (V, E)$ be a connected graph and all vertices have even degree. A chain that does not repeat an edge will eventually contain a circuit. We remove it. The degree of all vertices of each connected component of the resulting graph, again, is even. By induction it is a disjoint union of circuits.

Finally, assume G is connected and $E = C_1 \dot\cup \ldots \dot\cup C_k$ is a disjoint union of circuits. This time we use induction on k, the case $k = 1$ being trivial. Otherwise, each component of $G \setminus C_1$ is Eulerian by induction. Let the vertices of $C_1 = v_1, \ldots, v_l$ be numbered. As G is connected each such component contains a vertex of C_1 of smallest index and these disagree by pair. Our tour now follows C_1, including the Eulerian tour of the corresponding component each time we visit such a "contact vertex".   □

One possible algorithm to compute a Eulerian tour is due to Fleury. It is not implemented in CATBox and we give only a sketch.

Let $v_0 \in V$. Put T the empty word and start a chain in $v = v_0$ and iterate the following until E is empty:

- Choose an edge $e = (v, w)$ leaving v, such that the edge set $E \setminus e$ induces a connected graph that contains v_0.
- Put $E = E \setminus w$, $v = w$ and $T = Te$.

Theorem 37 *The algorithm of Fleury computes a Eulerian tour T.*

Proof The only thing we have to verify is that there is always a feasible edge. We proceed by induction on the steps of the algorithm. In the beginning such an edge clearly exists. Thus, assume that some feasible edge e_i has just been removed from E and that T and v have been updated accordingly. As the choice of e_i was feasible, the remaining edge set together with (v, v_0) admits a Eulerian tour. A choice of $e_{i+1} = (v, w)$ as an edge that follows (v_0, v) in some Euler tour clearly is feasible.
□

The crucial point in an implementation of Fleury's algorithm is the check for connectivity. It is easily done in $O(|E|)$ but that yields an overall bound of $O(|E|^2)$. On the other hand the proof of Theorem 36 is easily turned into an $O(|V| + |E|)$ algorithm.

Returning to the Chinese Postman Problem we see that the case where all vertices have even degree reduces to the Eulerian tour problem and thus is solved by either of the two algorithms mentioned above.

The general case reduces to the following:

Problem 12 Let $G = (V, E, w)$ be a connected graph with a non-negative weight function. Find a set of edges $F \subseteq E$ of minimum weight, such that replacing each edge of F with two edges makes G Eulerian.

Examining the necessary conditions for such an edge set F we see that a vertex v should have an odd degree in F if and only if it is odd in G. Vice versa doubling any edge set with that property yields a Eulerian graph. Following this we give a slight generalization of our problem:

Definition 34 Let $G = (V, E)$ be a graph and $T \subseteq V$ a subset of its edges. An edge set $F \subseteq E$ with the property that the degree $d_H(v)$ of a vertex in the graph $H = G[f]$ is odd if and only if $v \in T$ is called a *T-Join*.

Problem 13 Let $G = (V, E, w)$ be a graph with a non-negative weight function and $T \subseteq V$ a subset of its vertices such that $|T|$ is even. Find a T-join of minimum weight in E.

The problem becomes infeasible if T is odd:

Exercise 74 (Handshaking Lemma) Let $G = (V, E)$ be a graph. The number of vertices with odd degree is even.

Let us analyze the structure of T-joins.

Lemma 29 *Let F be a T-join in $G = (V, E, w)$. Then F decomposes into $\frac{|T|}{2}$ mutually edge disjoint paths between vertices from T and some circuits.*

Proof We proceed by induction on T. If $T = \emptyset$, the claim reduces to Theorem 36, thus let $|T| = 2k > 0$ and $t \in T$. According to the Handshaking Lemma the component of $G[F]$ containing T has to contain another odd vertex $s \in T$. Let P be an s, t-path in $G[F]$. Then $T \setminus P$ is a $\widetilde{T}$-join with $\widetilde{T} = T \setminus \{s, t\}$ and the claim follows by induction.   □

As we have non-negative weights the aforementioned paths in case of a minimum T-join have to be shortest paths. This yields a weighted matching on the complete graph K_T on the odd vertices, where the weights are given by the shortest path distances. Vice versa a minimum matching in this graph will consist of edge disjoint paths.

Lemma 30 *Let $|T| = 2k$ be a vertex set in a graph $G = (V, E)$ with strictly positive weights on the edges and dist : $T \times T \to \mathbb{R}$ the distance function (metric) for T in this graph. If M is a minimum matching in K_T weighted with dist, then the shortest paths between the matched vertices are mutually disjoint.*

Proof Assume P_i was an s_i, t_i-path for $i = 1, 2$ and $P_1 \cap P_2 \neq \emptyset$. Then $P_1 \triangle P_2$ contains two paths that change the pairing to a matching of smaller weight, a contradiction.   □

Summarizing:

Theorem 38 *An edge set F is a minimum T-join of a weighted graph $G = (V, E, w)$ for some vertex set $T \subseteq V$, if F decomposes into $\frac{|T|}{2}$ disjoint shortest paths, that induce a minimum weighted matching in K_T weighted with the distance function in G.*

The Chinese Postman Problem thus is solved as follows:

- Compute the distances for all pairs from $T \times T$, say with the algorithm of Floyd and Warshall.
- Compute a minimum weighted perfect matching in K_T weighted with these distances.
- Double the edges on shortest paths between matched vertices.
- Compute a Eulerian tour in the resulting Eulerian graph.

Exits

Matching Theory has become a field in its own in Discrete Mathematics. A good starting point for a deeper study and the classical reference is the book [31]. Also, half of Volume A of [38] is on matching theory and contains up to date results and references.

Exercises

Exercise 75 Prove that program (9.9) is half-integral, meaning that for any vertex v we have $2v \in \mathbb{Z}^{\binom{n}{2}}$.
Hint: Consider the $K_{n,n}$ and a map that maps two edges to each arc of K_n.

Exercise 76 Consider the complete graph in $V = \{1, 2, \ldots, n\}$ with weight function $c((i, j)) = i + j$. Perturb the weight function c as carefully as possible such that the path $1, 2, 3, 4, \ldots, n$ becomes a minimum spanning tree.

Exercise 77 Construct a Minimum-cost flow Problem that computes a minimum cost weighted perfect matching in a complete bipartite graph with non-negative weights.

Appendix A
Using Gato and Gred

CATBox consists of this book and a collection of graph algorithms and example graphs, see `http://schliep.org/CATBox` for more information, for which, just like for the book itself, Springer holds the copyright. The algorithms can be visualized using the software package Gato. Gato, which is copyright 1998–2009 Alexander Schliep and Winfried Hochstättler and copyright 1998–2001 ZAIK/ZPR, an institute at the Unversität zu Köln, is freely available under the Lesser General Public License (LGPL; see `http://www.gnu.org/licenses/lgpl.html`) from the website `http://gato.sf.net` where you can also find all the sources. The LGPL implies that you have a lot of freedom in using, learning from, making changes to and even redistributing Gato. If you have ideas how to improve Gato, which is just like the algorithms written in Python, and are willing to contribute, please contact either author. The software package Gato also contains a simple graph editor called Gred.

A.1 Easy Installation

For some popular operating systems we provide an easy way to install Gato and CATBox. Even if your operating system is not listed in this section you still can run CATBox. However, you have to install from source, see Sect. A.2, which is not that much harder.

We provide archive files which contain the CATBox algorithms and graphs and one binary executable each for Gato and Gred. Since we cannot rely on Python being available on Linux and Windows the executables therefore contain their own Python interpreter and are as a consequence rather large. However, the installation is pretty much just a couple of clicks with your mouse.

Apple MacOS X: Make sure that you are running MacOS X version 10.4 or above. Gato will not run on prior versions of Apple MacOS X and not on MacOS 9 or below. If you double-click the disk image `CATBox-Gato.dmg`, a disk called `CATBox-Gato` will be mounted and appear on your desktop or in the finder. Move the application `Gato` (the name will be shown in the Finder as `Gato.app` only if you switched on the display of file extensions) and the `Gred`-application to the

W. Hochstättler, A. Schliep, *CATBox*, DOI 10.1007/978-3-642-03822-8,
© Springer-Verlag Berlin Heidelberg 2010

system-wide `Applications`-folder or the one in your home-directory. Move the folder `CATBox` to a convenient location.

Linux: The archive `CATBox-Gato.tar.gz` contains all files required. Move it to the directory where you want to install CATBox and unpack the archive with the command `tar xvzf CATBox-Gato.tar.gz`. If you do not have GNU tar installed, then `gunzip < CATBox-Gato.tar.gz | tar xvf -` should work. Please make sure that `Gato.exe` and `Gred.exe` are in fact executable; you can move them elsewhere if you prefer. For some distributions you can also find ready-made packages for Gato; the algorithms and graphs you have to install from `CATBox.tar.gz` with the command `tar xvzf CATBox.tar.gz`. A Gentoo package can be obtained from `http://gentoo-portage.com/sci-misc/gato`. As Linux is a much more variable environment you might want to consider a installation from the sources, see Sect. A.2.

BSD: There is a port of Gato available under the name `py-gato`, see `http://portsmon.freebsd.org/portoverview.py?category=math&port name=py-gato` The algorithms and graphs you have to install from `CATBox.tar.gz` with the command `tar xvzf CATBox-Gato.tar.gz`.

Windows: The archive `CATBox.zip` contains all files required. Move it to the directory where you want to install CATBox and unzip it. Note, you might need to install third-party software to do so. You can move `Gato.exe` and `Gred.exe` elsewhere.

A.2 Installing from Source

When you are familiar with Python, the command line, know what a `tar`-file and what subversion is then you might want to install everything from source. This is not only guaranteed to work if your system can run Python and Tk, it also will give you the sources and an easy path to update to the newest versions of Gato.

Prerequisites: Please make sure that your Python is version 2.6.x. Currently, we do not support Python 3.x. You can check the version by typing `python` and return on your command line. On MacOS X there is `python` and `pythonw`; the latter you need to use for Tkinter applications unless you are running X11.

```
schliep@bayes> python2.6
Python 2.6.1 (r261:67515, Feb 24 2009, 14:31:46)
[GCC 4.2.0] on linux2
Type "help", "copyright", "credits" or "license" for more information.
>>> import Tkinter
>>> t = Tkinter.Tk()
>>>
```

The only other major package needed is Tcl/Tk, of version 8.4.7 or above and the Tkinter package for using Tk from Python. If everything is properly installed the command `t = Tkinter.Tk()` will produce a little window on your screen.

If one of the commands fails, you will need to install the package. You can download Python from `http://www.python.org/download/`. There you also will find Windows installer. The main Tcl/Tk-website is `http://www.tcl.tk/`. Almost all Linux distributions have binary packages for Python, Tkinter and Tk; check the package manager for your distribution.

On MacOS X Python and Tk is pre-installed and no further software is required. Note that if you use the pre-installed `pythonw` to start Gato the menu entry usually called Gato, next to the one with the apple symbol, will be called Python. You can install updates which also contain some further convenient Python modules from `http://www.python.org/download/`. Alternatively, if you want to run Gato under X11 then you might want to consider a separate Python installation for example using the Fink package manager from `http://fink.sourceforge.net/`.

Gato: There are three ways to obtain Gato: You can either use `Gato.tar.gz` the book website, download a snapshot from `http://gato.sourceforge.net/download.html` or checkout Gato anonymously from SVN with the following two commands:

```
svn co https://gato.sourceforge.net/svnroot/gato/trunk/Gato Gato
```

Gato uses the standard Python install mechanism using `setup.py`. With `python setup.py build` and `python setup.py install` you can install Gato in the default Python library directory. A different install location can be specified as `python setup.py install --prefix=/some/dir` which would install everything in `/some/dir/lib/python2.6/site-packages/`. This directory has to be added to your `PYTHONPATH`-environment variable. Note that the different directories are used for different major versions of Python. We found it convenient to place symbolic links to `Gato.py` and `Gred.py` into `/usr/local/bin`. Depending on your system and installation of Python it might be necessary to edit the first line (starting `#!`) of `Gato.py` and `Gred.py` to reflect the location of your Python executable. A quick test if everything works, is to start Python and enter **import** `Graph`. If there are problems with the installation this will produce an `ImportError`.

CATBox: Algorithms and graphs are contained in `CATBox.tar.gz`, which you should unpack in the directory in which you want to install the files using the command `tar xvzf CATBox-Gato.tar.gz`.

A.3 The Gato Application

The way to start Gato, the **G**raph **A**nimation **To**olbox, will depend on the operating system you use and how you installed Gato. On MacOS X it will be a double-click on the `Gato`-application, on Windows a double-click on `Gato.exe` and on Linux or Unix you will start Gato from the command line, for example by typing `gato`.

You first will see the splash screen (Fig. A.1) and then the two main windows of Gato, the algorithm window Fig. A.2, and the graph window Fig. A.4, will appear.

Fig. A.1 About Gato: License and information

At startup there will be neither algorithm nor graph. Either you can open from within Gato using the menu. On MacOS X the menu structure looks different to adhere to standards; you will find the same menu items, but in slightly different locations.

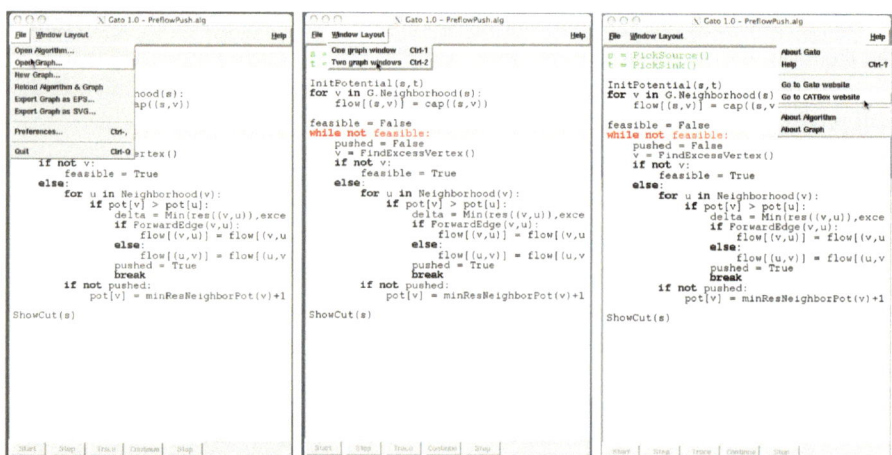

Fig. A.2 Algorithm window of Gato with the File (*left*), Window Layout (*middle*) and Help-menus (*right*)

On Unix, Linux and Windows the `File`-menu, Fig. A.2 (left), has the following entries.

`Open Algorithm...`: This menu item allows you to open an algorithm with default extension `alg` using a dialog for navigating directories and selecting files.

`Open Graph...`: This menu item allows you to open an Gred graph with default extension `cat` using a dialog for navigating directories and selecting files.

`New Graph...`: This starts the graph editor Gred. Note that this menu item is not available on MacOS X.

`Reload Algorithm & Graph`: This is a convenience function which allows you to quickly reload both the algorithm and the graph; this is useful, when you want to run an algorithm repeatedly on a graph you are editing in Gred.

`Export Graph as EPS...`: The complete contents of Gato's graph window is saved to a an Encapsulated Postscript File (EPS). This can be converted subsequently to other graphic formats. Useful for creating handouts or still images for presentations.

`Preferences...`: This brings up the `Preferences`-dialog, see Fig. A.3. There you can set the delay between instructions to artificially slow down

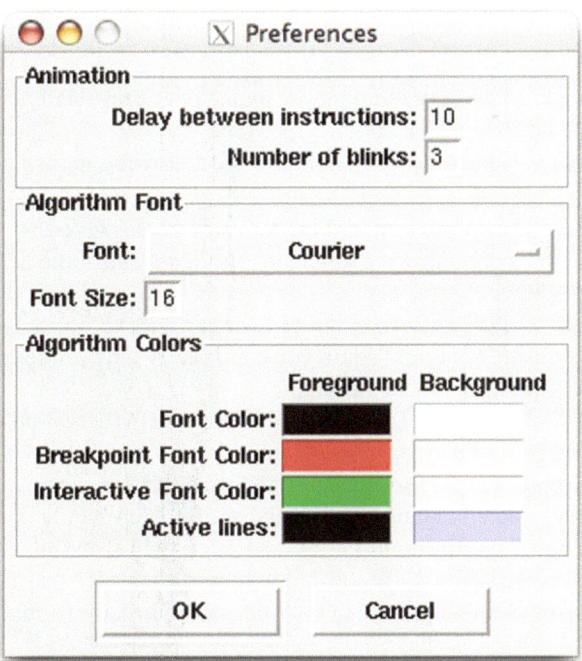

Fig. A.3 Gato's preferences dialog

algorithms and better observe what is happening. Similarly, the number of blinks determines how often an edge or vertex changes colors when the algorithm animation calls for blinking. You can set the font family and its size used for display of algorithms. The font style (normal, bold, or italic) is automatically determined to highlight the syntax of the Python code in the algorithm window. Lastly, you can set the color schemes. See Appendix. A.3.1 for details of the meaning of the different colors. The preferences are stored per user in a file called .gato.cfg in the user's home directory.

Quit: This quits the Gato application.

The Window Layout-menu, Fig. A.2 (middle), provides functionality to arrange windows on the screen. Some algorithms need to display two graphs simultaneously, for example the Ford-Fulkerson algorithm for maximal flows in capacitated networks.

One Graph Window: The algorithm window will be moved to the left- and top-most position. Its size remains unchanged. One graph window will be moved to the right of it and its size will be changed to fill up the remaining screen.

Two Graph Windows: The algorithm window will be moved to the left- and top-most position. Its size remains unchanged. The remaining space will be used for two graph windows which will be stacked onto each other with their sizes changed to fill up the remaining screen.

Note that on Linux a few very old window managers occasionally do not properly function with the underlying Tk code.

The Help-menu, Fig. A.2 (right), contains the following items.

About Gato: The splash screen, Fig. A.1, will be displayed.

Help: A window containing the keyboard shortcuts and some help information will be shown.

Go to Gato Website: An external browser will be opened displaying the Gato website. On Unix/Linux you can specify the browser used with your BROWSER environment variable.

Go to CATBox Website: An external browser will be opened displaying the CATBox website.

About Algorithm: Algorithms provide some further information about their purpose, implementation and visualization. Here you can read about the meaning of the different vertex and edge colors and other visualization cues you see while the algorithm is running.

About Graph: Some graphs also provide some further information.

The graph window of Gato displays the graph for which different fixed zoom-levels can be chosen; see Fig. A.4. At the bottom of the graph window next to the

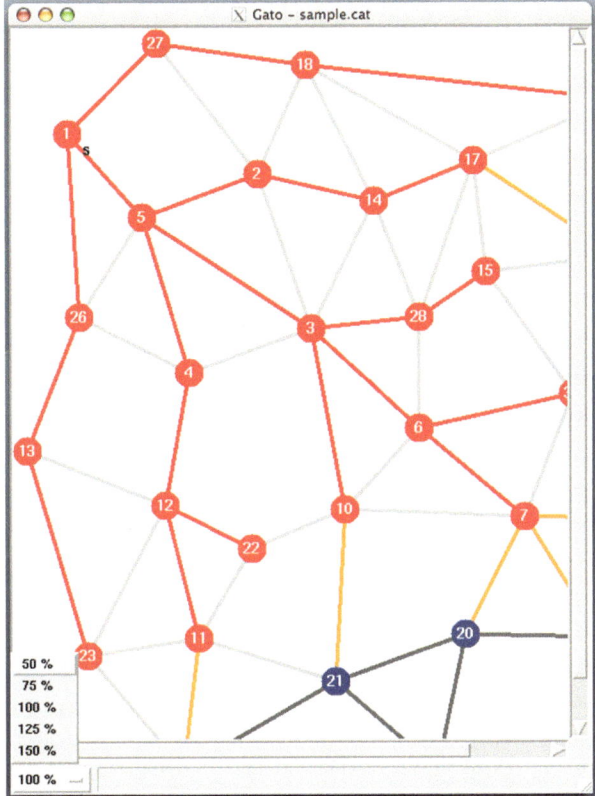

Fig. A.4 The graph window in Gato allows to display the graph at different zoom-levels

zoom you find an info box where additional information about the algorithm's state or about vertices and edges, if you place the mouse pointer over them, is displayed.

A.3.1 Running Algorithms in Gato

Before you can run an algorithm you have to open the algorithm and a graph to run it on. You can see that once you do, the state of the Start-button at the bottom of the algorithm window changes, see Fig. A.5 (left). The font turns from grey, or deactivated, to black, or activated, and you can press the Start-button. Once you do, the other buttons change their state to active too, Fig. A.5 (right), and the Start-button becomes deactivated again.

The actions controlled by the buttons are as follows:

Start: Start execution of an algorithm. Clicking Start activates the other buttons and deactivates the Start-button. Also, the display of the active line (see below) indicates that an algorithm is being executed.

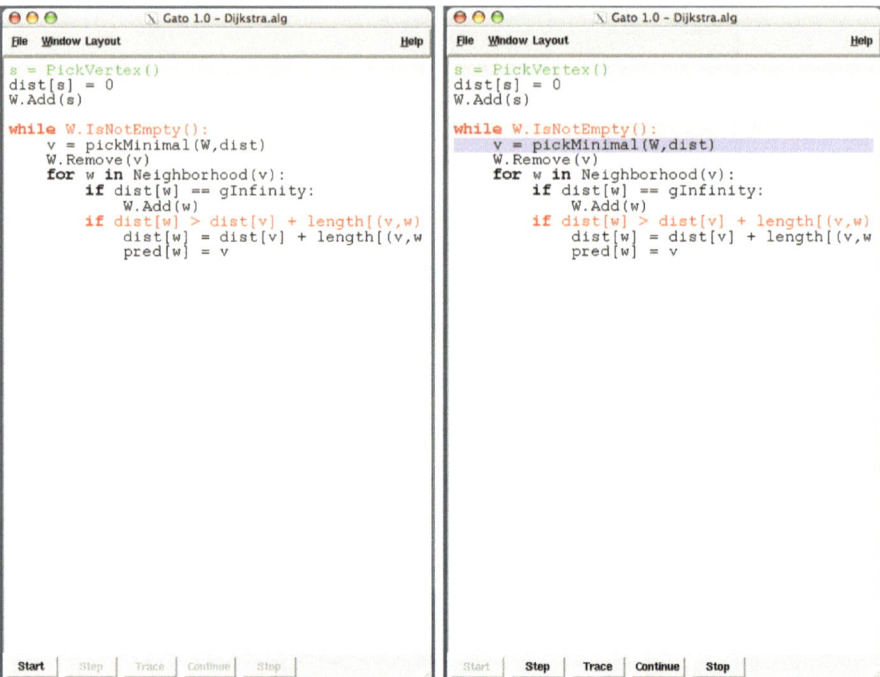

Fig. A.5 States of the algorithm window. Once an algorithm and a graph have been opened, the `Start`-button becomes active (*left*). Once the algorithm has started the other buttons become active (*right*) and remain so, until the algorithm is stopped or terminates

`Step:` Switch into step mode; continue execution by executing the algorithm until it arrives at another line of the algorithm.

`Trace:` Step into a function being called. Only works for function defined in the same algorithm file. Continues execution by executing the algorithm until it arrives at the first line of the function called.

`Continue:` Switch into run mode; continues execution.

`Stop:` Terminate execution of the algorithm.

You will notice that the algorithm code has syntax highlighting; Python keywords for example are shown in bold face. Also, you will notice different colors. As the actual colors used are defined in the `Preferences`-dialog we will describe the defaults here; you will be able to distinguish different foreground and background color combinations corresponding to different types of lines in Fig. A.5 (right).

The active line: The line which will be executed next is displayed with a light blue background and black type. The active line helps you to follow the execution path as the algorithm is running.

Breakpoint lines: Often it is helpful to stop run-mode at specific lines of an algorithm. This we will call a breakpoint. Breakpoints are displayed with a red font and a light grey background. Breakpoints are predefined for some algorithms or can be added by clicking on the line. In both cases you can remove a breakpoint by clicking on that line. If a line with a breakpoint becomes active only its background color will change.

Interactive lines: Sometimes algorithms ask the user for input such as choosing a vertex or an edge by clicking on it; the functions called `PickVertex()` or `PickEdge()` are examples for such interaction. A line requiring user interaction is called an interactive line and is usually displayed with a green font on a light grey background. If an interactive line becomes active only its background color will change. If you click either `Step` or `Continue` some vertex or edge will be selected for you. If you choose step, another step is necessary to proceed to the next line.

Normal lines: These are all other lines of the algorithm usually displayed with black font on white background.

Note that there are keyboard short-cuts available for all relevant commands. Unlike the menu short-cuts, which you can read off the menu entries and which differ between operating systems, the algorithm commands are triggered from the keyboard without any modifier keys such as control, command or windows. Check the help in Gato for an up-to-date list.

Key	Command
s	Start
space	Step
t	Trace
c	Continue
x	Stop
b	Toggle breakpoint in active line

A.4 Gred

Gred, the **Graph Ed**itor, is a simple application for creating graph instances for use in Gato.

The `File`-menu, Fig. A.6 (top left), contains the following items. On MacOS X you find the `Quit`-item in a different menu to adhere to standards.

New: Start editing a new graph.

Open...: This menu item allows you to open an Gred graph with default extension `cat` using a dialog for navigating directories and selecting files.

Save: Save the graph being edited.

Save as...: Save a copy of the graph being edited into a file in a location you can specify using a dialog for navigating directories and selecting files.

Export EPSF...: The complete contents of Gred's graph window is saved to a an Encapsulated Postscript File (EPS). This can be converted sub-

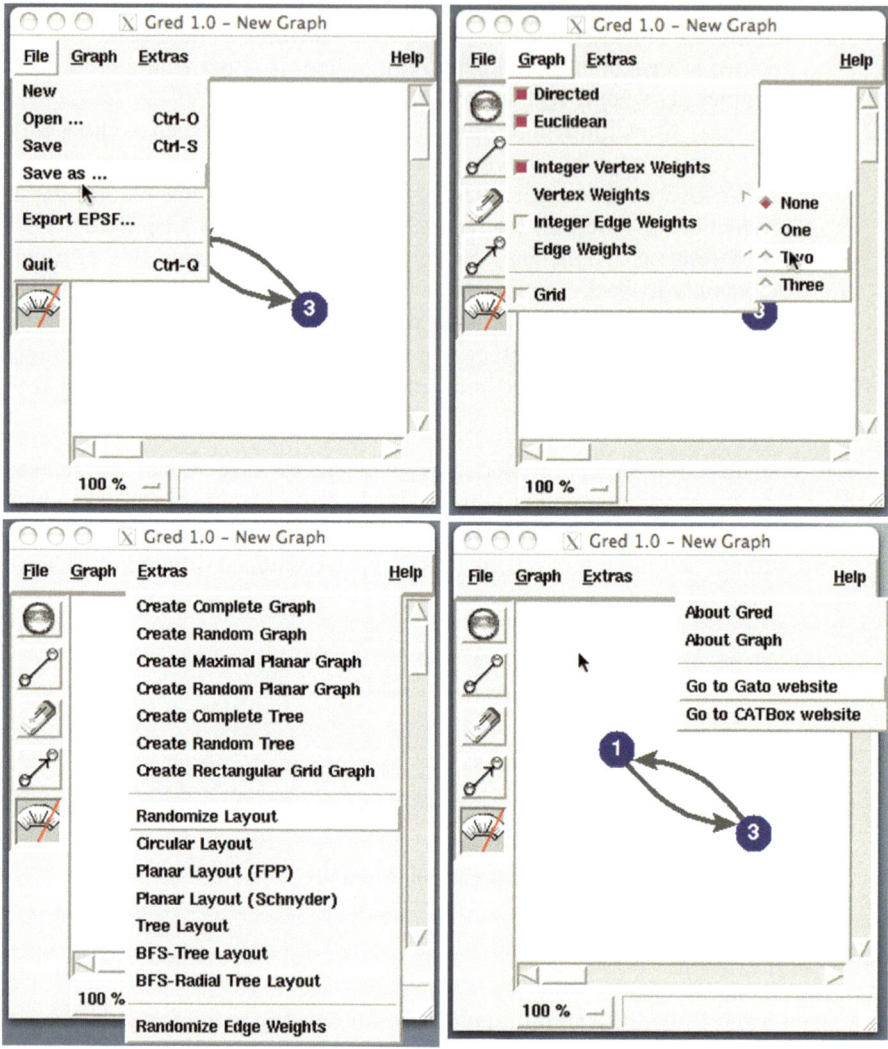

Fig. A.6 The Gred window with the different menu items. On the *left* you see the vertical toolbar with different tools for editing graphs

sequently to other graphic formats. Useful for creating hand-outs or still images for presentations.

Quit: This quits the Gred application.

The Graph-menu, Fig. A.6 (top right), allow you to define properties of the graph you are editing.

Directed: This toggles between a directed, or digraph, and an undirected graph.

Euclidean: If this is selected, it is assumed that the graph is embedded into the usual, Euclidean, two-dimensional space and the weight of an edge is determined by the Euclidean distance between incident vertices. If off, you can set the weights arbitrarily.

Integer Vertex Weights: This chooses between integer and real-valued vertex weights.

Vertex Weights: With the sub-menu you can choose the number of vertex weights for vertices in the graph.

Integer Edge Weights: This chooses between integer and real-valued edge weights.

Edge Weights: With the sub-menu you can choose the number of edge weights for edges in the graph.

Grid: If switched on, you cannot place or move vertices arbitrarily in the editor. Rather the permissible positions are constrained to a square grid of size 50 pixel by 50 pixel.

The Extras-menu, Fig. A.6 (bottom left), allows you to create certain types of graphs automatically and also provides some simple functions for computing certain kinds of embeddings, or placements of vertices.

Create Complete Graph: Create a complete simple graph, where every vertex is adjacent to every other vertex except itself. You can specify the number of vertices, whether the graph should be directed or undirected, and the layout on nodes on a circle or randomly in the plane.

Create Random Graph: Create a random graph on a specified number of vertices and edges. All edge sets of the selected cardinality have equal probability. You can also specify whether the graph should be directed or undirected, and the layout on nodes on a circle or randomly in the plane.

Create Maximal Planar Graph: A planar graph can be drawn in the usual Euclidean two-dimensional space without crossing edges. You can specify the number of vertices; Gred will add the maximal number of edges possible while maintaining planarity. You can specify whether the graph should be directed or undirected and select a layout algorithm.

Create Random Planar Graph: You can choose the number of vertices and edges. All edge sets of the selected cardinality are have equal probability. You can also specify whether the graph should be directed or undirected and select a layout algorithm.

Create Complete Tree: You can specify the degree of each non-leaf vertex in the tree and the height, that is the distance from the root to any leaf. You can also choose whether the graph should be directed or undirected and select a layout algorithm.

Create Random Tree: This creates a random induced sub-tree of a complete tree with a specified number of vertices.

Create Rectangular Grid Graph: This creates n by m vertices spaced on a square grid with edges connecting horizontal and vertical neighbors.

Randomize Layout: Vertices are placed independently and uniformly on the canvas at random.

Circular Layout: Vertices are placed every $360/n$ degrees, where n is the order of the graph, onto a circle.

Planar Layout (FPP): If the graph is planar, compute a layout with the FPP-algorithm [15].

Planar Layout (Schnyder): If the graph is planar, compute a layout with the Schnyder's algorithm [37].

Tree Layout: If the graph is a tree, compute a rooted tree layout from a specified root.

BFS-Tree Layout: Independent on whether the graph is a tree, compute a BFS from a specified root and place vertices in a tree-like layout according to distance to the root (or BFS-level). This layout uses small horizontal offsets to allow display of edges connecting vertices on the same level.

BFS-Radial Tree Layout: If the graph is a tree, compute a rooted tree layout from a specified root in a radial fashion. This works well for trees where the non-leave vertices have a high degree.

Randomize Edge Weights: Choose the edge weights randomly. Note that selecting Euclidean will override any weights edited upon saving the graph.

The Help-menu, Fig. A.6 (bottom right), contains only the About Gred-item, which displays Gred's about-box.

A.4.1 Editing Graphs with Gred

Editing of graphs works with different tools which you switch between from the toolbar on the left of the main Gred window, see Fig. A.7. You can, just as in Gato, choose different zoom levels to display the graph, see Fig. A.7, and modify it, using different tools. At the bottom of window next to the zoom menu you find an info box where additional information about vertices and edges is displayed, when you place the mouse pointer over them. The tools for editing are, from top to bottom in the toolbar, the following:

Add or Move Vertex tool: Pressing and releasing the mouse button on the canvas in an empty spot will produce a new vertex. If you press over an existing vertex you can move the vertex by moving or dragging the mouse. The vertex will remain at the position when you release the mouse button. Note that you cannot choose positions arbitrary if the grid option is selected in the Graph-menu.

Add edge tool: Pressing the mouse button over an existing vertex and dragging it to another vertex and releasing it, will create an edge or an arc connecting the two.

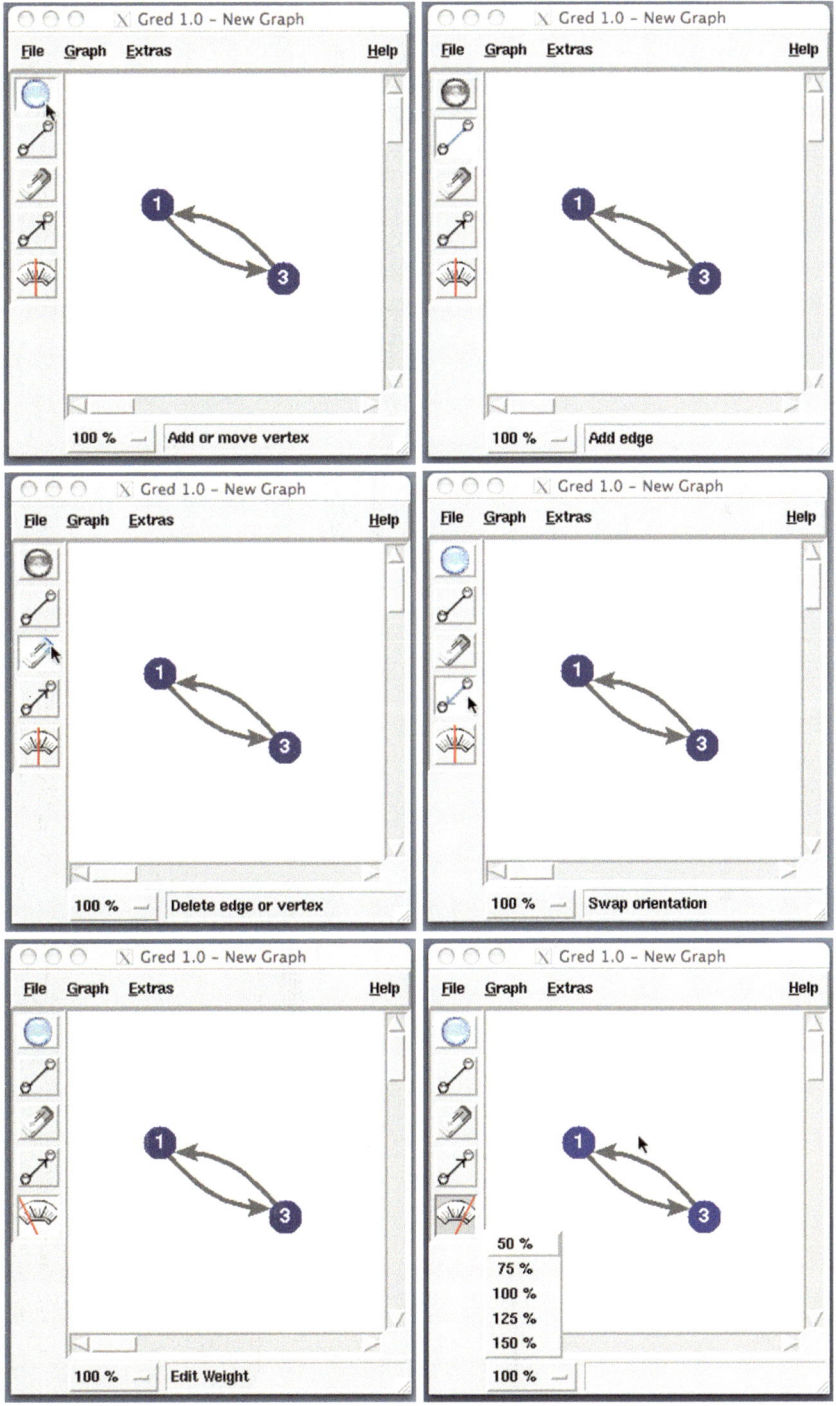

Fig. A.7 The Gred window showing different editing modes, which are selected from the toolbar on the *left*, *Bottom right*: Gred can also display the graph at different zoom-levels

Delete edge or vertex tool: Releasing the mouse button over an existing vertex or
    edge will delete it. Deleting vertices also deletes all incident edges.

Swap orientation tool: In a directed graph you can swap tail and head of an arc.

Edit weight tool: Clicking on edges or vertices will bring up an dialog to edit
    weights, if you have specified that the graph has vertex or edge weights in
    the Graph-menu.

If your mouse has three or more buttons, you can do most editing of graphs with
the *Add or Move Vertex*-tool. The left mouse button will add or move vertices, the
middle mouse button will add edges by clicking and dragging and the right mouse
button will delete edges or vertices.

Appendix B
A Brief Introduction to Reading Python

Python is a procedural language, it is dynamically typed and object-oriented and has gained a large amount of attention and use over the last ten years. The best-known company to use Python on a very large scale is Google; in fact, Google currently employs Guido von Rossum, Python's inventor, and many other major contributors from the Python community. However, our choice of Python a couple of years back was not motivated by Python's success in the enterprise but rather by its simplicity and beauty.

```
def factorial(n):
    if n == 0:
        return 1
    else:
        return n * factorial(n-1)
```

While the code resembles other procedural languages like C, Pascal or Java, one immediately notices the absence of block delimiters like {...}, semi-colons, or BEGIN and END statements. In Python, the indentation not only helps the reader to understand the code; it is significant and denotes the block structure. One consequence is that badly formatted code will simply not run. The **def**-statement designates that the following code is the *definition* of a function called factorial. Note that no prior declaration of argument and return types is necessary. The function takes one *argument*, called n; it is up to the user of the function that the argument is of an appropriate type, at least allowing multiplication and subtraction.

The body of the function—the subsequent four lines which are indented one level—consists of an **if**-then-**else** block. Note the use of the colon to separate the actions from the conditions. If n equals zero, a single "=" would mean assignment, the function returns one, else the recursive formula yields a *return value* of n * factorial(n-1), so we call the factorial function recursively.

If the code example is saved in a Python-file called factorial.py you can compute factorials with the interactive Python interpreter from a command line as follows.

```
schliep@karlin> python
Python 2.3.3 (#1, Mar 16 2004, 11:20:51)
[GCC 2.95.3 20010315 (release)] on linux2
Type "help", "copyright", "credits" or "license" for more information.
>>> from factorial import *
>>> factorial(12)
479001600
```

First, we start Python on the command line with the command `python`, some information is printed and the Python prompt >>> appears; here you can enter arbitrary Python commands, for example **print** `"Hello World"`. The **import** statement tells Python to load all symbols—functions, classes and variables—from the file `factorial.py` so that we can call the factorial function.

If we want to create a *list* of the first 10 factorials we can simply write the sequence $0, \ldots, 9$ in square brackets and assign them to the variable r. More convenient is the use of the `range`-function to obtain the same list. Then a *list comprehension*—a concise way of creating lists—creates a new list by calling `factorial()` for every number in the list r.

```
>>> r = [0, 1, 2, 3, 4, 5, 6, 7, 8, 9]
[0, 1, 2, 3, 4, 5, 6, 7, 8, 9]
>>> print range(10)
[0, 1, 2, 3, 4, 5, 6, 7, 8, 9]
>>> [factorial(n) for n in r]
[1, 1, 2, 6, 24, 120, 720, 5040, 40320, 362880]
```

What you have seen in the list comprehension, **for** n **in** r (line 5 above), is the pythonic way of applying a function to all values in a container. Loops in general execute the same block of code for different values; more often than not, the different values are taken from some container like in this example. A container is anything which holds values like lists, arrays, trees, dictionaries, queues, stacks; here the container is the list r.

Algorithms in Gato have access to the underlying graph data structure and some further convenient variables. The Python variable `Vertices` for example is a list of all the vertices in the graph you loaded into Gato. An iteration over all vertices can be done by the command **for** `root` **in** `Vertices:`, which is actually the third line of the `BFS-components.alg` in Chap. 1.

ALGORITHM BFS-components

```
step = 1

for root in Vertices:
    if not label[root]:
        component[root] = NewComponent()
        component[root].AddVertex(root)
        Q.Append(root)
        pred[root] = root

        while Q.IsNotEmpty():
            v = Q.Top()
            label[v] = step
            step = step + 1
            for w in Neighborhood(v):
                if not pred[w]:
                    component[v].AddVertex(w)
                    component[w] = component[v]
                    Q.Append(w)
                    pred[w] = v
```

There are two further aspects of Python we have not seen before. First, we need to store some information, namely the predecessor or the vertex from which we found v. We use another container, a dictionary which provides a mapping between a *key* and a *value*, to keep track of this. The dictionary is called `pred`, and for the key w the value can be retrieved with `pred[w]` (see the last line of `BFS-components.alg`) and set to v with `pred[w]` = v. It is a more flexible alternative to arrays you might be familiar with from other programming languages where the *index*, which we call a key, has to be an integer and `somearray[i]` is similarly used for getting and setting values. Actually, Python lists can be used in this way. An empty Python dictionary is created with `pred` ={}.

Second, we actually use objects here; for example on lines 5, 6 and 7 of `BFS-components.alg`. Objects are a convenient way of combining data and functions operating on it. The goal is to encapsulate functionality and thus make the implementation and interface more independent. Let us use the queue Q as an example. A queue is a data structure where the first value in the queue can be retrieved, a function typically called `Top()`, and further values can be appended at the end, `Append()`. Think of the checkout line at a supermarket. The implementation of the queue needs to store the values which are appended somewhere and needs to provide functionality to retrieve them. How this is done in detail is of no concern to us when we want to use a queue. Moreover, it might be advantageous if we do not rely on the peculiarities of the implementations, but rather write code which works with arbitrary queue implementations. All we need to know is how to create a queue, how to append values and how to retrieve them.

Functions associated with objects are called methods and the way to call them is through the object using the dot-notation: the `Append()`-method of the queue Q is called as `Q.Append(v)` and the `Top()`-method is called as `Q.Top()`. Objects are created as instances of classes; a class is the recipe describing the object and instance creation or instantiation is the process of creating an object in memory. Let us assume we had two queue classes providing different implementations, `SpeedyQueue` and `NaiveQueue`, then we could instantiate a queue object Q either as Q = `NaiveQueue()` or as Q = `SpeedyQueue()`. Our code would not know the difference when using the queue object, Q[1].

Gato and its classes provide a rich functionality. For graph objects created with G = `Graph()` (you need to import `Graph.py` for this to work outside Gato) methods are provided to add and delete vertices and edges, e.g. `G.AddVertex()`, to obtain order, `G.Order()`, and size, `G.Size()`. The encapsulation into a graph object makes sure that consistency of the data structure is maintained when adding and deleting vertices or edges.

B.1 Where to Obtain Python

The official Python website is `http://www.python.org` and there one finds the Python software, its sources and its complete documentation. Also, links to community projects and software packages written in Python are provided. Many

---

[1] Note that the Python standard started to provide more abstract data types recently, which followed a different naming convention.

modern operating systems, for example MacOS X and several Linux distributions, already come with Python pre-installed.

Exits—Further Python Resources

There is a tutorial at `http://docs.python.org/tut/tut.html` which is aimed at readers with some computing background. There is also a large number of Python books available; which one appeals to you is hard to predict, so browsing online or in a real bookstore might be a good option to choose one. As Python is actively used in introductory programming classes from middle school to university, quite a lot of very nice introductory material is available online and there is even a special interest group for using Python in education, `http://www.python.org/community/sigs/current/edu-sig/`. For example the online textbook "How to Think Like a Computer Scientist Learning with Python" by Allen B. Downey, Jeffrey Elkner and Chris Meyers, see `http://www.ibiblio.org/obp/thinkCSpy/`, provides a real nice introduction to complete novices. Further alternatives are listed at `http://wiki.python.org/moin/BeginnersGuide/NonProgrammers`. Dive into Python, a book for experienced programmers by Mark Pilgrim is available at `http://www.diveintopython.org/`.

Appendix C
Visualizing Graph Algorithms with Gato

There is a long history of visualizing or animating algorithms and complete software systems. In particular graph algorithms, excuse the pun, lend themselves to graphic representations or visualizations. For many mathematicians and computer scientists a visualization of the workings of an algorithm is something drawn on paper or on a blackboard. Sometimes different stages of an algorithm make it as a sequence of figures into textbooks. The lack of interactivity—how does a shortest path look like in another graph is a question which cannot be answered by a book—and the economical constraints imposed by printing color figures, motivate the use of computers to provide alternatives. There have been a large number of systems visualizing algorithms, which have been used for teaching on a surprisingly large-scale basis even in the early eighties; see references given in the exits of Appendix C for a glimpse of the history.

The visual language, the visualization cues used, was quite obvious and quite universal for the various visualization systems. Vertices and edges blink, they change colors, their width or their shape, they are added or deleted, they move their position, or they are annotated with additional information such as labels. What really distinguishes the different visualization systems is how the visualizations are created for an algorithm as this determines cost and time the development process takes. The question of how to link algorithm and visualization, of linking *cause*—a specific statement of the algorithm—with *effect*—the resulting visual cue—is closely related, and for didactic purposes even more important.

C.1 Interesting Events During Runtime

Independent of the technicalities of implementing a visualization, we first need to decide what we want to visualize: Which specific statements of an algorithm should be brought to the attention of the learner? In the literature, the concept of *interesting events* [7] was proposed, which basically are crucial steps in algorithms. We will use the breadth first search (BFS) algorithm from Chap. 2, which you find in the directory 02-GraphsNetworks, as an example.

ALGORITHM BFS

```
   root = PickVertex()
   Q.Append(root)
   pred[root] = root
   step = 1
5
   while Q.IsNotEmpty():
       v = Q.Top()
       label[v] = step
       step = step + 1
10     for w in Neighborhood(v):
           if not pred[w]:
               Q.Append(w)
               pred[w] = v
```

The algorithm is straight-forward and consists of a few simple steps. The body of the **while**-loop can be summarized as *processing* an unlabeled vertex v; hence one interesting event is processing v. Clearly, it would be helpful, if the visualization would indicate which of the vertices displayed is v. Another interesting event is the *exploration of the neighborhood* of v. With the statement **for** w **in** Neighborhood(v): we iterate over all neighbors w of v, checking whether w is labeled or not. Labeled w are not that interesting, however whenever we find an unlabeled w this is another interesting event. We say that we *visit* w when we encounter an unlabeled w while processing v. Visited w have to be processed later on, so we store them in the queue. In summary, we would like to visualize the events processing v, exploring neighbors w, and visiting w. You find the following three algorithms in the directory Appendix.

ALGORITHM BFS-Visualization-1

```
   root = PickVertex()
   Q.Append(root)
   pred[root] = root
   step = 1
5
   while Q.IsNotEmpty():

       v = Q.Top()
       A.SetVertexColor(v,'red')
10
       if not label[v]:
           label[v] = step
           step = step + 1
15         for w in Neighborhood(v):
               A.BlinkEdge(v,w)
               if not pred[w]:
                   A.SetVertexColor(w,'blue')
                   Q.Append(w)
20                 pred[w] = v
```

In Gato, the animator object can be referred to as A. The animator provides primitives for drawing and editing graphs and a wide range of visual cues you can use for building algorithm visualizations. Here, we only change the color of a vertex using the SetVertexColor()-method to visualize the event of processing v by changing its color to red, A.SetVertexColor(v, 'red'). Similarly, when we visit w we change its color to blue, A.SetVertexColor(w, 'blue'). The exploration of neighbors is a transient event and so we just let the edge (v, w) blink. Compare the two visualizations of `BFS.alg` and `BFS-Visualization-1.alg`.

Not all that necessary, but informative and helpful in understanding the algorithm is the display of values of variables used. In `BFS.alg` we use the `pred`-dictionary to memorize the predecessor of w as `pred[w]`. Predecessor means that we visited w while processing v. This implies that (v, w) is an edge in the graph (or an arc in the case of a digraph) and a natural way of visualizing the value of `pred [w]` is to color the edge `(pred[w],w)` in a signal color. Note that the `pred`-dictionary is where we store the BFS-tree we compute. By visualizing the edges we will see the complete BFS-tree once the algorithm terminates. The BFS-labeling `label[v]`, which indicates the order in which we processed vertices, can be displayed numerically.

ALGORITHM BFS-Visualization-2

```
root = PickVertex()
Q.Append(root)
pred[root] = root
step = 1

while Q.IsNotEmpty():

    v = Q.Top()
    A.SetVertexColor(v,'red')

    if not label[v]:
        label[v] = step
        A.SetVertexAnnotation(v,label[v])
        step = step + 1

        for w in Neighborhood(v):
            A.BlinkEdge(v,w)
            if not pred[w]:
                A.SetVertexColor(w,'blue')
                Q.Append(w)
                pred[w] = v
                A.SetEdgeColor(v,w,'red')
```

So we add two more visualization commands: We change the edge color when we set `pred[w]` and we show a vertex annotation when we set the label. If you compare `BFS.alg` and `BFS-Visualization-2.alg` closely, you will observe that it is still harder to track the vertex currently being processed. One possibility is to highlight it by changing the width of its outline (or frame width in Tk parlance) using `A.SetVertexFrameWidth(v,6)`. However you need to keep track of the previous vertex whose frame width you changed, to change it back to the default.

C.2 Rule-Based Visualization

Adding visualization commands to algorithms more complicated than a BFS is tedious, cumbersome and prone to errors and inconsistencies. It would be helpful if one did not have to indirectly define visualization of interesting events by inserting visualization commands wherever such an interesting event occurs in the algorithm. Rather, the definition of rules—show visualization X when interesting event Y occurs—linking cause and effect would be desirable.

While we can easily specify rules in natural language, for example *highlight vertex being processed*, it is a little bit more complicated to formalize it for a programming language. Events can be tied to one or several lines of code, but almost always relate to something implicit to the code. For many algorithms, interesting events can be defined differently and indirectly. In the BFS for example, there are four categories of vertices and an interesting event always occurs, when one vertex changes from one category to the next.

The four categories are

(i)   Vertices which have not been visited, yet.
(ii) Vertices which have been visited, but not processed.
(iii) The vertex being processed.
(iv) Vertices which have been processed.

If you go back to BFS-Visualization-2.alg and inspect placement and effect of the visualization instruction, you will observe that we were in fact already visualizing categories 1, 2, and 4. That is, we were visualizing the event of a vertex changing from one category to another, as those correspond exactly to our interesting events.

Breaking the algorithm into phases and dividing vertices or edges or subgraphs into different category is a general approach for thinking about algorithms and their visualizations. From our experience, this is virtually always possible. However, the enumeration does not provide technical means to implement rule-based visualizations. We will propose one implementation in the next section.

A very important argument in case of rule-based visualization is the following: Observe that in BFS-Visualization-2.alg the instructions of the algorithm, the very steps a student is actually trying to understand, are obfuscated by the visualization instructions. Some visualization systems provided technical tricks around that, but when you execute a visualized algorithm, one will always have to mix algorithm and visualization instructions. This makes it very hard to use the algorithm implementation which is actually being executed for teaching.

A further consequence of the rule-based approach is the possibility of changing algorithms while maintaining the visualization, opening the possibility of students implementing algorithm variants. Moreover, semi-automated creation of, at least rough, algorithm visualizations is certainly possible.

C.3 Animated Data Structures (ADS)

The four categories of vertices we defined above can be reformulated as changes to the core underlying data structure we use in the BFS. This enables us the formulate the visualization rules in a manner which we can easily implement and which is transparent to the algorithm implementation. The queue is the data structure which keeps the order of vertices explored. Initially, all the vertices are in the first category, which corresponds to vertices which are neither currently nor have been previously been on the queue. Visiting a vertex entails appending it to the queue, so the vertices in the queue belong to the second category. The BFS processes the first vertex in line. It is the only vertex in the third category and it changes to the fourth category if the next vertex in line is processed.

Category	Status
1. Vertices neither visited nor processed	Vertices never in queue
2. Visited Vertices	In queue
3. Vertex being processed	Last vertex removed from queue
4. Processed Vertices	Removed from queue

The interesting events we define as a vertex changing from one category to another can now be easily visualized by changing the queue implementation from a general data structure to an animated one, in this case, to a `AnimatedVertexQueue`. The `Queue`-class has two main methods, `Top()` and `Append()`. In our `Animated VertexQueue` we will change the implementation of the `Append()`-method to also take care of the visualization we added to `BFS-Visualization- 2.alg`. That is, we add the visualization command `A.SetVertexColor(w, 'blue')` to the `Append()`-method, see Fig. C.1. In the `Top()`-method we add the `A.SetVertexColor(v, 'red')` visualization command. There, we can also add the change the frame width of the vertex being processed as proposed in Sect. C.1. We can use an instance variable of the `AnimatedVertexQueue`-class, `self.lastRemoved` to remember which vertex was last returned by `Top()`, so that we can change its frame width back to the default and change the frame width of the vertex we will return using `A.SetVertexFrameWidth(v,6)`. In summary, we define the following interesting events for vertices changing categories and implement them based on the queue.

Category change	Implementation
1. → 2.	Vertex appended to queue: `Q.Append()`
2. → 3.	Vertex removed from queue: `Q.Top()`
3. → 4.	Another vertex removed from queue: `Q.Top()`

Through the ADSs the visualization of an algorithm becomes transparent, it becomes consistent and it becomes much easier to produce than through the individual visualization commands which we added to `BFS.alg` in Sec. C.1. Be warned though, that for more complex algorithms, the definition of vertex and edge categories and algorithm phases becomes harder, as does the definition of suitable ADSs.

Software Exercise 78 There is a BFS implementation without any visualization called `BFS-NoVisualization.alg` in the directory `Appendix`. Open the corresponding prolog, `BFS-NoVisualization.pro`, in an editor and locate the `Q = Queue()` command. Change that line to `AnimatedVertexQueue (A, 'blue', 'red')`. The additional argument tell the `AnimatedVertex-Queue` to use the animator A and the colors it should use. Compare `BFS-No Visualization.pro` with `BFS-NoVisualization.pro` in the directory `02-GraphsNetworks` and identify further animated data structures used.

C.3.1 Case Study: Ford-Fulkerson Algorithm

The Ford-Fulkerson algorithm, see Sect. 6.2, computes a maximal flow in a capacitated network. The core idea of the algorithm is quite simple. One finds a shortest source-sink path in a residual network, which initially is a copy of the capacitated network input. Along the shortest path we can send an amount of flow determined by the minimum residual capacity of the arcs on the path. We record the amount of flow in the network and subtract the flow from the residual capacity in the residual network. The only non-obvious idea is that one has to insert backward arcs (w, v) in the residual network if there is a positive flow along the arc (v, w). Also, zero-capacity edges are removed from the residual network.

ALGORITHM FordFulkerson

```
   def UpdateFlow(Path):
       delta = MinResCap(Path)
       for (u,v) in Edges(Path):
           if ForwardEdge(u,v):
 5             flow[(u,v)] = flow[(u,v)] + delta
           else:
               flow[(v,u)] = flow[(v,u)] - delta

   s = PickSource()
10 t = PickSink()

   while not maximal:
       Path = ShortestPath(s,t)
       if Path:
15         UpdateFlow(Path)
       else:
           maximal = true

   ShowCut(s)
```

The two main data structures used are dictionaries for storing the flow and the residual network. With respect to the flow value edges can be in different categories; `cap` and `res` designate original capacity and residual capacity respectively.

Flow value	Network	Residual network
`flow[(v,w)] == 0`	Unused	Full capacity forward
`0 < flow[(v,w)] < cap[(v,w)]`	Used	Backward arc added
`flow[(v,w)] == cap[(v,w)]`	Capacitated	Forward arc removed
`res[(v,w)]` minimal	Minimal	

One subset of interesting events to visualize pertains to the search for a shortest path in the residual network using a variant of the BFS algorithm, see Sect. 2.5. This is visualized like the BFS algorithm in the previous section and will not be discussed here; you find the animated algorithm `shortestPath` in the module `AnimatedAlgorithms`.

That an arc is the arc—more correctly, one of the arcs—determining the maximal increase in flow along a shortest path is not a property of the `flow` data structure per se, but rather a property of the shortest path computed in relation to the flow. Hence, the visualization of the fourth, transient category *arc determines maximal increase in flow* is done within the function `MinResCap` computing this maximal permissible increase using the `SetEdgeColor()` method. The interesting events with respect to the flow for the first three categories are the following.

Category change	Visualization
1. → 2.	Flow increases: edge color changes, add backward arc
2. → 3.	Arc capacitated: edge color changes, remove forward arc
3. → 2.	Flow decreases: edge color changes, add forward arc
2. → 1.	Flow decreases to zero: edge color changes, remove backward arc

As the same visualization also applies to the Preflow-Push algorithm, see Sect. 6.7, there is a ADS called `FlowWrapper` which implements the functionality. It keeps track of original capacity, flow and residual capacity and checks for interesting events, whenever the flow is changed using `flow[(v,w)] = value`. Edge colors are changed using the `SetEdgeColor()` method of the animator using distinct colors for unused, used and capacitated edges. Also the current flow and capacity values are displayed numerically using the `SetEdgeAnnotation()` method. Adding and removing arcs is done using the `AddEdge()` and `Delete Edge()` methods.

Finally, one minimal cut is displayed as a proof of optimality coloring the edges of the cut in a signal color.

If you inspect the prolog `FordFulkerson.pro` you will find code for setting up the two graph displays, copying the input network to the residual network, choosing sink and source etc. Also, there we define code for displaying sensible information when one moves the mouse over vertices or edges.

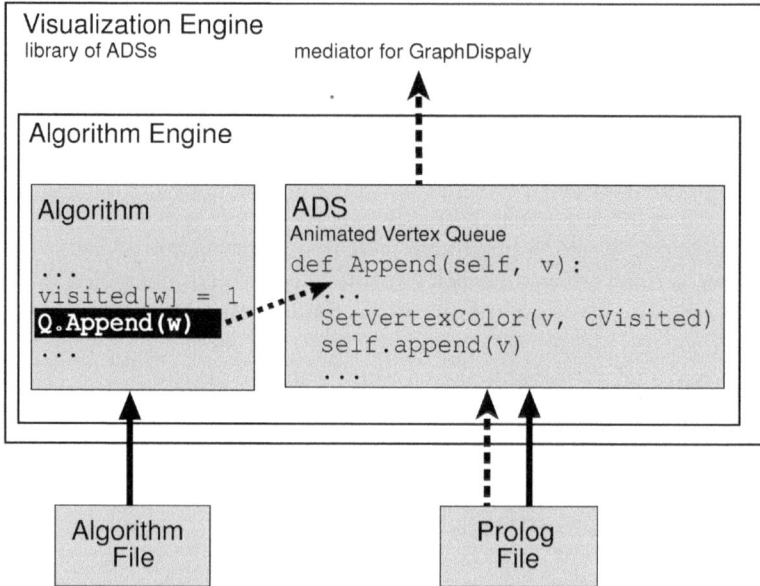

Fig. C.1 Here we depict one crucial step and the components involved in the animation of the graph algorithm `BFS.alg` using a queue

C.3.2 Implementation of Gato

The software architecture of Gato has a few major components; see Fig. C.2: the algorithm engine, the visualization engine, the algorithm display and the graph display. As Gato is free open-source software, you can inspect the implementation in detail. There is also documentation available for developers, which is derived from the comments in the sources; see `http://gato.sourceforge.net/Documentation/Internal/index.html`.

Input files: When one loads an algorithm into Gato, actually two files are read, `algorithm.alg` which contains the executable Python code and `algorithm.pro`, the prolog. The prolog also contains Python code for providing ADS which are not part of Gato proper, setting variables such as default breakpoints and the algorithm information displayed by the `About Algorithm` menu item. The prolog can also provide the necessary initial setup and the definition of functions. The text from `algorithm.alg` is passed on to the algorithm display. Similarly, graphs are read and passed on to the graph display.

Algorithm engine: The algorithm engine runs algorithms in a controlled fashion, allowing single line stepping, tracing into functions and stopping execution. It is controlled from the graphical user interface of the algorithm display and calls methods of the algorithm display to provide feedback of its status. The algorithm engine is implemented as a subclass of the default Python debugger; it makes use of the fact that Python is interpreted and that it is thus trivial to execute arbitrary Python code dynamically during run-time. The quasi-concurrency of algorithm execution and handling events in the graphical user interface is realized not in execution threads but rather by use of the event processing of the Tk graphical user interface toolkit.

Visualization engine: The ADS and methods in the graph display form the visualization engine which provides visual feedback for algorithms which are executed. It is already possible to exchange the graph display used, as the engines and displays are clearly separated.

Algorithm display: The algorithm display uses the Tk graphical user interface toolkit through the Tkinter Python package. The major component is a Tk text widget which is used for the display of algorithms.

Graph display: The major component of the graph display is the Tk canvas, which already provides object-oriented vector drawing abilities. Objects such as the circles used for displaying vertices and lines which are used for edges can be modified in appearance, hidden or deleted. The use of Tk canvas features allows to provide the rich visualizations in Gato. Methods of the graph display provide a convenient interface for displaying, editing and changing the visual appearance of Graphs. The graph editor Gred is actually a subclass providing a GUI for the `GraphDisplay` -class. The Graph class and its subclasses implemented in Gato are abstract graphs with an additional embedding into Euclidean space.

C.3.3 ADS Defined in Gato

Most animated data structures are defined in the Python file `AnimatedData-` `Structures.py` and listed in the following. However, for some algorithms in CATBox you will find ADSs defined in the prolog files, as this made development easier—changed definitions of ADSs can be tested without quitting and restarting Gato. Here we give an overview over general capabilities; you find a more detailed reference to the different ADSs and all their options at `http://gato.source` `forge.net/Documentation/Internal/AnimatedDataStructures.` `py.html`. Sometimes you will find alternative animated data structure which allow the same type of visualization; some ADSs are tied to very specific algorithm types.

C.3.3.1 Iterating over Neighbors

AnimatedNeighborhood Visualize iterating over the neighbors w of a vertex v in a graph G by changing the color of traversed edges (v,w). Replace calls

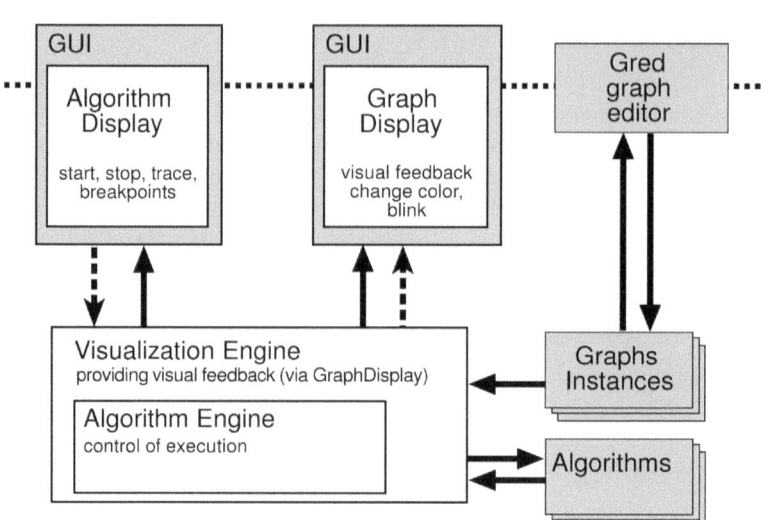

Fig. C.2 The fundamental building blocks in Gato

G.Neighborhood(v) with AnimatedNeighborhood(A,G,v). Can be used just like a list.

BlinkingNeighborhood Visualize iterating over the neighbors w of a vertex v in a graph G by letting traversed edges (v,w) blink. Replace calls G. Neighborhood(v) with BlinkingNeighborhood(A,G,v). Can be used just like a list.

BlinkingTrackLastNeighborhood Visualize iterating over the neighbors w of a vertex v in a graph G by letting traversed edges (v,w) blink and temporarily changing the color of the last traversed edge to grey. See Blinking Neighborhood for usage. Can be used just like a list.

C.3.3.2 Containers

BlinkingContainerWrapper Visualize iterating over lists of edges or vertices by blinking. Replace references to a list l with references to al for al = BlinkingContainerWrapper(A,l,'red'). Can be used just like a list.

ContainerWrapper Visualize iterating over lists of edges or vertices by changing their color. Once the color has been changed, further references such as $l[i]$ to a vertex or edge will not change the color again, if it has been modified externally. Replace references to a list l with references to al for al = Container Wrapper(A,l,'red'). Can be used just like a list.

AnimatedVertexQueue Provide a vertex queue and visualizes vertices with respect to the queue. User can specify colors used for vertices on the queue and removed from the queue;AnimatedVertexQueue(A,colorOn, color Removed. Provides methods Top() and Append().

AnimatedVertexPriorityQueue Similar to AnimatedVertexQueue but provides a priority queue with methods Insert(), DecreaseKey() and Delete Min()

AnimatedVertexStack Similar to AnimatedVertexQueue but provides a stack with methods Push() and Pop(). Can be used like a set.

AnimatedVertexSet Similar to ContainerWrapper but using a Python set structure. You can define the color used for vertices when they are removed from the set in AnimatedVertexSet(A,set,color). Can be used like a set.

AnimatedEdgeSet Similar to ContainerWrapper but using a Python set structure. You can define the color used for edges removed from the set in Animated EdgeSet(A,set,color). Can be used like a set.

C.3.3.3 Labelings

VisibleVertexLabeling Visualize a vertex labeling with strings usually displayed to the bottom right of a vertex. If you define al = VisibleVertex Labeling(A) then al[v] = 5 will display the value 5 next to vertex v. If you pass values gInfinity or –gInfinity then the strings 'Infinity' respectively '-Infinity' will be displayed.

AnimatedVertexLabeling Similar to VisibleVertexLabeling. Additionally, changes in value will cause vertices to change color. In Animated Vertex Labeling(A, initial, color) you can specify a designated initial value, vertices will be colored cInitial and the color for visualizing changes.

BlinkingVertexLabeling Similar to VisibleVertexLabeling. Vertices blink as the the labeling is changed.

C.3.3.4 Subgraphs

AnimatedPredecessor Visualize a predecessor or parent array pred by coloring edges (pred[v],v]) red and grey if the value pred[v] is subsequently changed. In instantiation with AnimatedPredecessor(A,leave Colors) one can specify a list of colors not to change.

AnimatedSubGraph Provides an induced sub-graph of G which visualizes membership by changing colors of its element vertices and edges. Instantiation with AnimatedSubGraph(G,A,color).

C.3.3.5 Problem-Specific ADS

FlowWrapper Provides visualization for maximal flow algorithms. Given a directed, capacitated network G and a directed capacitated residual network R changes to the flow variable will result in updates to the residual capacities, adding and deleting of edges in the residual network, change of edge annotations in the network and change of colors to highlight saturated edges. Usage `flow = Flow Wrapper(G,A,R,RA,G.edgeWeights[0],R.edgeWeights[0])` where `RA` is the animator for the residual network. Optionally one can specify an additional array used for storing the excess flow in a vertex, which will also be visualized.

ReducedCostsWrapper Provides visualization of the reduced costs of an edge by coloring edges with positive reduced costs green, with zero costs grey and with negative costs red. Instantiation with `rc = ReducedCostsWrapper(A, cost,pot)`. Can be used like a dictionary; i.e. `rc[(1,4)] = 5`.

Exits

The authorative work and survey giving an overview about the development of the field from the early 1980s [6] is the book Software Visualization [40]. It also gives accounts of the large-scale implementation and use of algorithm animation at Brown university; also see [41, 24]. Since then, a wide range of packages and methods have appeared, some of which use a declarative approach to defining animation [12, 30]. Some current developments are reflected in two proceedings volumes [5, 17]. You can find links to systems for example at `http://www-cg-hci.informatik. uni-oldenburg.de/~da/peters/Kalvin/Doku-TN.htm` (in German, but you will find the links).

References

1. R. AHUJA, T. MAGNANTI, AND J. ORLIN, *Network Flows*, Prentice Hall, 1993.
2. R. E. BELLMAN, *Dynamic Programming*, Dover Publications, Incorporated, 2003.
3. D. P. BERTSEKAS AND J. N. TSITSIKLIS, *An analysis of stochastic shortest path problems*, Math. Oper. Res., 16 (1991), pp. 580–595.
4. O. BORŮVKA, *O jistém problému minimálnim*, Práce mor. přírodověd. spod. Brně, III (1926), pp. 37–58.
5. J. BORWEIN, M. H. MORALES, K. POLTHIER, AND J. F. RODRIGUES, eds., *Multimedia Tools for Communicating Mathematics*, Springer, Heidelberg, 2002.
6. M. H. BROWN AND R. SEDGEWICK, *A system for algorithm animation*, in Proceedings of ACM SIGGRAPH '84, July 1984, pp. 177–186.
7. M. H. BROWN AND R. SEDGEWICK *Interesting events*, in Software Visualization. Programming as a Multimedia Experience, J. Stasko, J. Domingue, M. H. Brown, and B. A. Price, eds., MIT Press, Cambridge, MA, 1998.
8. F. R. K. CHUNG, *Spectral Graph Theory*, CBMS Regional Conference Series in Mathematics, American Mathematical Society, 1997.
9. V. CHVATAL, *Linear Programming*, W. H. Freeman, New York, Sept. 1983.
10. W. COOK, W. CUNNINGHAM, W. PULLEYBLANK, AND A. SCHRIJVER, *Combinatorial Optimization*, Series in Discrete Mathematics and Optimization, Wiley-Interscience Publications, New York, 1998.
11. T. CORMEN, C. LEISERSON, AND R. RIVEST, *Introduction to Algorithms*, The MIT Electrical Engineering and Computer Science Series, MIT Press, Cambridge, MA, 1994.
12. P. CRESCENZI, C. DEMETRESCU, I. FINOCCHI, AND R. PETRESCHI, *LEONARDO: A software visualization system*, in Proceedings of the 1-st Workshop on Algorithm Engineering (WAE'97), 1997, pp. 146–155.
13. E. DAHLHAUS, D. S. JOHNSON, C. H. PAPADIMITRIOU, P. D. SEYMOUR, AND M. YANNAKAKIS, *The complexity of multiterminal cuts*, SIAM J. Comput., 23 (1994), pp. 864–894.
14. G. DANTZIG, *Linear Programming and Extensions*, Princeton University Press, Aug. 1998.
15. H. DE FRAYSSEIX, J. PACH, AND R. POLLACK, *Small sets supporting fary embeddings of planar graphs*, in STOC '88: Proceedings of the Twentieth Annual ACM Symposium on Theory of Computing, New York, NY, USA, 1988, ACM, pp. 426–433.
16. A. DELCHER, S. KASIF, R. FLEISCHMANN, J. PETERSON, O. WHITE, AND S. SALZBERG, *Alignment of whole genomes*, Nucl. Acids Res., 27 (1999), pp. 2369–2376.
17. S. DIEHL, ed., *Software Visualization. International Seminar Dagstuhl Castle*, Springer, Heidelberg, 2002.
18. R. DIESTEL, *Graph Theory*, Graduate Texts in Mathematics, Springer, Berlin, 1979.
19. J. H. EATON AND L. A. ZADEH, *Optimal pursuit strategies in discrete state probabilistic systems*, Trans. ASME Ser. D, J. Basic Eng., 84 (1962), pp. 23–29.
20. S. EVEN AND H. GAZIT, *Updating distances in dynamic graphs*, Meth. Oper. Res., 49 (1985), pp. 371–387.

21. M. R. GAREY AND D. S. JOHNSON, *Computers and Intractability: A Guide to the Theory of NP-Completeness*, Freeman, San Francisco, 1979.

22. O. GOLDSCHMIDT AND D. S. HOCHBAUM, *Polynomial algorithm for the k-cut problem*, in FOCS, IEEE, 1988, pp. 444–451.

23. P. HART, N. NILSSON, AND B. RAPHAEL, *A formal basis for the heuristic determination of minimum cost paths*, Systems Science and Cybernetics, IEEE Transactions on, 4 (July 1968), pp. 100–107.

24. S. E. HUDSON AND J. T. STASKO, *Animation support in a user interface toolkit: Flexible, robust and reusable abstractions*, in Proceedings of the 1993 ACM Symposium on User Interface Software and Technology, Atlanta, GA, ACM, Nov. 1993, pp. 57–67.

25. V. JARNIK, *O jistém problému minimálnim*, Práce mor. přírodověd. spod. Brně, VI (1930), pp. 57–63.

26. M. JÜNGER AND W. PULLEYBLANK, *Geometric duality and combinatorial optimization*, in Jahrbuch Überblicke Mathematik, S. Chatterji, B. Fuchssteiner, and U. K. R.Liedl, eds., vol. 111, Vieweg, 1993, pp. 1–24.

27. R. KANNAN, S. VEMPALA, AND A. VETTA, *On clusterings—good, bad and spectral*, in FOCS, 2000, pp. 367–377.

28. B. KORTE AND J. VYGEN, *Combinatorial Optimization: Theory and Algorithms*, Springer, Berlin, 4th ed., Oct. 2007.

29. J. KRUSKAL, *On the shortest spanning subtree of a graph and the travelling salesman problem*, Proc. Amer. Math. Soc. 7 (1956), pp. 48–50.

30. LEONARDO: *A C programming environment for reversible execution and software visualization*. URL http://www.dis.uniroma1.it/ demetres/Leonardo/.

31. L. LOVASZ AND M. D. PLUMMER, *Matching Theory*, 1st ed., Elsevier Science Ltd, Amsterdam, Aug. 1986.

32. R. MOTWANI AND P. RAGHAVAN, *Randomized Algorithms*, Cambridge University Press, New York, NY, Aug. 1995.

33. J. NEŠETŘIL, *A few remarks on the history of MST-problem*, Archivum Mathematikum (Brno), 33 (1997), pp. 15–22.

34. L. RABINER, *A tutorial on hidden markov models and selected applications in speech recognition*, Proceedings of the IEEE, 77 (Feb 1989), pp. 257–286.

35. L. RODITTY AND U. ZWICK, *On dynamic shortest paths problems*, in ESA, S. Albers and T. Radzik, eds., vol. 3221 of Lecture Notes in Computer Science, Springer, 2004, pp. 580–591.

36. S. RUSSELL AND P. NORVIG, *Artificial Intelligence: A Modern Approach (2nd Edition)*, Prentice Hall, London, Dec. 2002.

37. W. SCHNYDER, *Embedding planar graphs on the grid*, in SODA '90: Proceedings of the first annual ACM-SIAM symposium on Discrete algorithms, Philadelphia, PA, USA, 1990, Society for Industrial and Applied Mathematics, pp. 138–148.

38. A. SCHRIJVER, *Theory of Linear and Integer Programming*, Wiley-Interscience Publications, New York, NY 1986.

39. A. SCHRIJVER, *Combinatorial Optimization*, Springer, Feb. 2003.

40. J. STASKO, J. DOMINGUE, M. H. BROWN, AND B. A. PRICE, eds., *Software Visualization. Programming as a Multimedia Experience*, The MIT Press, Cambridge, MA, 1998.

41. J. T. STASKO, *The TANGO algorithm animation system*, Tech. Report CS-88-20, Department of Computer Science, Brown University, Dec. 1988.

42. R. J. VANDERBEI, *Linear Programming: Foundations and Extensions*, Springer, 3rd ed., Nov. 2007.

43. A. VITERBI, *Error bounds for convolutional codes and an asymptotically optimum decoding algorithm*, Information Theory, IEEE Trans., 13 (Apr 1967), pp. 260–269.

44. M. S. WATERMAN, *Efficient sequence alignment algorithms*, J. Theor. Biol., 108 (1984), pp. 333–7.

45. J. ZHANG, S. XU, AND Z. MA, *An algorithm for inverse minimum spanning tree problem*, Optim. Methods Softw., 8 (1997), pp. 69–84.

Index